A "Without Formulae" Book

FLIGHT WITHOUT FORMULAE

HOW AND WHY AN AEROPLANE FLIES
EXPLAINED IN SIMPLE LANGUAGE

A. C. KERMODE
C.B.E., M.A., F.R.Ae.S.

Author of
"Mechanics of Flight" and "The Aeroplane Structure,"
and co-author of "Hydrofoils"

FOURTH EDITION

PITMAN

PITMAN BOOKS LIMITED
128 Long Acre, London WC2E 9AN

Associated Companies
Pitman Publishing Pty Ltd, Melbourne
Pitman Publishing New Zealand Ltd, Wellington
Copp Clark Pitman Ltd, Toronto

© A. C. Kermode 1970

Fourth edition published in Great Britain 1970
Reprinted 1973, 1975, 1977, 1978, 1979, 1982

All rights reserved. No part of this publication may be reproduced, stored in a retrieval system, or transmitted, in any form or by any means, electronic, mechanical, photocopying, recording and/or otherwise without the prior written permission of the publishers. This book is sold subject to the Standard Conditions of Sale of Net Books and may not be resold in the UK below the net price.

Printed and bound in Great Britain
at The Pitman Press, Bath

ISBN 0 273 40360 5

PREFACE TO FOURTH EDITION

This little book, in its original form, was the outcome of various criticisms of a book on the same subject which I wrote some few years previously. One reviewer indicated that it was "not at all bad, except that it was mixed up with so much mathematics". Others were a little kinder; but I feel no resentment against any of them, because they did show me that there was a demand for a book on the subject without any mathematics whatsoever, and thus they encouraged me to try my hand again and, although it was more difficult than the other one, it was also more interesting and amusing. Moreover, the success of the earlier editions has shown the need for non-mathematical books on other aeronautical subjects, and so a series of such books has been built up—all "without formulae", but all treating a technical subject seriously.

To forestall any criticism of the opposite nature, let me make it quite clear to the mathematically minded reader—and reviewer—that this book is not meant for him. If he is one of those—and there are many—who can see no sense in taking two pages to explain something which can be expressed much more clearly in one line by a mathematical formula, then let him look elsewhere for information. There are plenty of books to which he can turn, but there are not so many for those others for whose benefit I have written. I have tried to give them as much up-to-date and correct knowledge of how and why an aeroplane flies as is possible without the use of mathematics. The reader cannot expect to learn everything in this way; but if, at the end of it all, he feels that he does not know very much about it, let him console himself with the thought that the mathematical student does not know much more. That is the way with this subject.

The book is unorthodox in many ways. For this I make no apology so long as the reader does not feel that he has been cheated. He will, if he is expecting an ordinary textbook. All I have done is to ramble on from paragraph to paragraph about the various problems that crop up when the subject is discussed. I have written as the mood has taken me. The first and second persons have been freely used in a deliberate attempt to depart from the textbook atmosphere. If some part of the subject has seemed interesting, if it is commonly misunderstood or argued about, then more space has been devoted to it than it properly deserves. Other parts may have been dismissed rather too lightly, perhaps because they seemed dull, or simply because they raise no argument. Much may seem simple, and it undoubtedly is; but let the reader beware of what is too "obvious"—because in this subject it is usually wrong!

Nearly thirty years have passed since the book was first published. It is hardly surprising that in some respects it had become out of date—and that in spite of a very thorough revision in preparing the third edition some eight years ago. It is difficult to believe, for instance, that in the first edition there was no mention of jet propulsion or of flight at supersonic speeds. My first task, therefore, in preparing new editions has been to fill such obvious gaps. But that was not enough. As I read through the book, not once but many times, I realized that much more than this was necessary if I was to be fair to my readers. During this time the very language of flight has changed; words and phrases that once sounded reasonable enough have now a slightly old-fashioned air, if indeed they are in current use at all; hopes or fears of the future have become realities of the present, or have been relegated to the past; photographs of what was then the latest type of aircraft remind one of the family album. So what might have seemed a small task became a big one; every sentence in the book has

PREFACE

been considered, and many have been altered; the words, the headings, the diagrams, the photographs, all have been examined critically to decide whether or not they were in keeping with a book of today. In this edition, as in the last and in other books in the series, the photographs have been put at the end; this saves breaking the continuity of the text, but, even more important, it makes it possible to round off a story told in words with one told in pictures. Only the author—and publishers—can possibly know what has been involved in the revision, but they will be well rewarded if the reader is satisfied, and saved the irritation of reading what is obviously out of date. Of course the process will continue, and this edition in its turn will become old-fashioned; but that is inevitable, and the reader will surely recognize that the author cannot do more than write of what he knows at the present.

Since the last edition was published it has been decided that Britain shall adopt the metric system of units. Fortunately in a non-mathematical book of this kind the change makes very little difference, since any figures that are given are mainly for comparison one with another, and the comparisons are not affected by the system of units. However, and particularly since the aviation industry are pioneering the change, a table of the relative values of the few quantities used in this book is given on page xiii.

Perhaps it is not without significance that, in preparing an early edition, the bulk of the revision was done by the author while actually in the air—on a flight of nearly twenty thousand miles from the Far East to the United Kingdom and back. When one sits at a desk and writes about flying, one sometimes wonders whether the things that one writes about really do happen, and gazing out of the window brings neither confirmation nor inspiration; but when, sitting in an aeroplane, seeing the clouds rush past—always coming from the front—feeling the extra lift when the flaps go down, gazing at an aileron which remains depressed even when the wings are level, then one writes about flight as it really is, flight without embellishments, flight without formulae.

In preparing the later editions I have not been so fortunate and I have had to do most of the work on the ground; but in spite of this handicap I have tried to write from an understanding of the point of view of the pilot and crew of the aircraft, and perhaps even of the aircraft itself, because to those of us with a lifetime interest in flying an aircraft seems to display very individual, and almost human, characteristics.

My thanks are due to many: to the aircraft firms who have co-operated by providing photographs; to the publishers; and perhaps most of all to those to whom I have tried to teach this subject. They run into many thousands, and whatever they may or may not have learnt from me, I feel that each one has taught me some little thing, even if it was only by asking some question to which I did not know the answer.

A. C. K.

1969

CONTENTS

In order to preserve continuity of the argument, the usual method of dividing a book into chapters, each covering a different aspect of the subject, has been avoided. For reference purposes, however, the main sections have been given headings, and they are also numbered. A complete list of section headings is given below. In the index at the end of the book the references are to page numbers.

	PAGE
Preface	iii

Section

1.	The Argument	1
2.	What is an Aeroplane?	1
3.	Lighter than Air	3
4.	Lighter than Air—more Problems	10
5.	The Atmosphere	12
6.	Lift and Drag	17
7.	Air Speed and Ground Speed	19
8.	Direction Relative to the Air and Relative to the Ground	22
9.	Wind Tunnels	23
10.	Smoke Tunnels	28
11.	Air and Water	29
12.	Centre of Pressure	31
13.	Stability and Instability	32
14.	The Wing Section	34
15.	Air Flow over a Wing Section	35
16.	Pressure Distribution Round a Wing Section	37
17.	The Venturi Tube	40
18.	Why the Centre of Pressure Moves	45
19.	Stalling or Burbling	46

Section	PAGE
20. Lift and Drag again	49
21. Effects of Speed	50
22. Effects of Size	51
23. Effects of Air Density	53
24. Lift/Drag Ratio	54
25. Analysis of Drag	55
26. Induced Drag	57
27. Parasite Drag	60
28. Form Drag	62
29. Skin Friction	65
30. The Boundary Layer	67
31. Shape of Wing Section	72
32. Variable Camber	73
33. Slots, Slats and Flaps	74
34. Aspects Ratio	77
35. Biplanes	80
36. Lift and Drag—A Summary	84
37. Straight and Level Flight	85
38. The Four Forces	86
39. Thrust	88
40. Jet Propulsion	89
41. Propeller Propulsion	90
42. Rocket Propulsion	92
43. Balance of Aeroplane	94
44. The Tail Plane	98
45. Stability of Aeroplane	100
46. Degrees of Stability	103
47. Rolling, Pitching, and Yawing	105
48. Longitudinal Stability	106
49. Lateral Stability	108
50. Directional Stability	110
51. Directional and Lateral	111
52. Control	112

CONTENTS

Section	PAGE
53. Longitudinal Control	114
54. Lateral Control	114
55. Directional Control	115
56. Balanced Controls	116
57. Control Tabs	119
58. Control at Low Speeds	122
59. Control at High Speeds	127
60. Level Flight—The Speed Range	131
61. Economical Flying	134
62. Flying at Low Speeds	137
63. Stalling	137
64. Landing	139
65. Reduction of Landing Speed	143
66. Wing Loading	145
67. S.T.O.L. and V.T.O.L.	146
68. Gliding	150
69. Climbing	163
70. Turning	171
71. Nose-Diving	180
72. Taxying	183
73. Taking Off	184
74. Aerobatics	186
75. The Propeller	196
76. Multi-Engined Aeroplanes	205
77. Flying Faults	206
78. Instruments	213
79. The Air-Speed Indicator	215
80. The Altimeter	218
81. Navigation Instruments	220
82. Flight Instruments	223
83. High-Speed Flight	226
84. The Speed of Sound	226
85. Mach Numbers	229

CONTENTS

Section	PAGE
86. Flight at Transonic Speeds	231
87. Shock Waves	232
88. The Shock Stall	232
89. Wave Drag	235
90. Sweepback	238
91. Vortex Generators	240
92. Wing and Body Shapes	242
93. Through the Barrier—and Beyond	243
94. Supersonic Flow	247
95. Supersonic Shapes	248
96. Sonic Bangs	251
97. Other Problems of Supersonic Flight	252
98. The Future	255
99. Into Space	256
100. Happy Landings!	263
The Final Test	264
Index	272
List of Plates	276

FLIGHT WITHOUT FORMULAE

1. The Argument

I am going to try to explain how an aeroplane flies. This does not mean that I am going to teach you how to fly an aeroplane—that is a very different matter. Many people who can explain how an aeroplane flies cannot fly one. Still more can fly an aeroplane, but do not know how it flies. A few people can do both.

Now, if you ask brainy people to explain to you how an aeroplane flies, they will tell you that it is all very complicated. If you persist in your search for knowledge they will instruct you by means of formulae, Greek letters, and various kinds of mathematics. When you are thoroughly fogged, they will shake their heads sadly and tell you that your knowledge of mathematics is insufficient to tackle the rather advanced problems involved in the flight of an aeroplane.

Mind you, there is some truth in what they say. If you wish to be an aeronautical professor, or a designer of aeroplanes, you must, sooner or later, acquire a fair knowledge of mathematics. But I take it that you have not got any such ambitions, at any rate for the present, and that you will be content with a simple explanation of the main principles on which the flight of an aeroplane depends.

That is all I am going to give you; and that is why I have called this book *Flight Without Formulae*.

2. What is an Aeroplane?

If you look up the definition of an aeroplane in a glossary, you will find that it is described in some such terms as these:

"A heavier-than-air flying machine, supported by aerofoils, designed to obtain, when driven through the air at an angle inclined to the direction of motion, a reaction from the air approximately at right angles to their surfaces."

There's a mouthful for you! When you have finished reading this book, you may care to look at this definition again. If you do so, you will find that it is perfectly sound and is a rather clever attempt to put a large amount of information into a few words. That is the object of a definition, and that is why a glossary makes rather dull reading in spite of the care which has often been exercised to ensure that conciseness should not lead to misunderstanding.

Many aeronautical books either begin or end with a glossary; but I prefer to explain any terms which may be necessary as and when we come across them. Even when explanation is necessary, the use of a hackneyed definition will be avoided because I want you to understand the term rather than learn to repeat, like a parrot, a string of technical words.

What, then, is an aeroplane?

All man-made contrivances which fly, that is to say which are kept in the air by forces produced by the air, are called *aircraft*.

There are two main kinds of aircraft: those which are *lighter than air* and those which are *heavier than air*. The former include *airships*, *balloons*, and *captive* or *kite balloons*; these are supported in the air not, as is commonly supposed, by the gas inside them, but rather by the air which this gas displaces. It is not the purpose of this book to deal with this type of aircraft, but a brief summary of the principles of their flight will be given. The latter, or heavier-than-air type, consists of many different forms which can conveniently be grouped under two headings, *power-driven* and *non-power-driven*—to which we should perhaps add a third, the very interesting *man-power-driven* (one of the problems of flight

WHAT IS AN AEROPLANE?

that is still only on the threshold of being solved). The non-power-driven forms are *gliders*, *sail planes* and *kites*.

The distinction between a glider and a sailplane is a subtle one, the latter being a lighter type which is able to "soar" in up-currents of wind. Every boy knows what a kite is, so I will not trouble to explain it. It might be imagined that, in these days, every boy knows what an aeroplane is, but unfortunately there has been much confusion over the terms used for heavier-than-air power-driven aircraft.

In an attempt to minimize the confusion, the British Standard Glossary of Aeronautical Terms divides them into three types—*aeroplanes*, *rotorcraft* and *ornithopters*. The term *aeroplane* includes aircraft which fly off the land and those which fly off the water, and, of course, *amphibians*, which can fly off either. This means that a *seaplane* is merely a particular type of aeroplane so designed as to be able to fly off and on to water, and therefore, to distinguish them, aeroplanes which can only fly from land are classified as *land planes*. Seaplanes themselves may be divided into two types, *float planes* and *flying boats*.

It will be noticed that *helicopters*, and other types of rotary-wing aircraft—the distinction between the three types will be explained later—are, strictly speaking, not aeroplanes at all; nor is the flapping-wing ornithopter, though that won't worry us very much. Whether *hovercraft* are a form of aircraft is still disputable.

Fig. 1 and the photographs at the end of the book should help to make the various terms clear. Fig. 2 shows the names of some of the main parts of a land plane; if you are not already familiar with them have a look at them now, they will help you to understand the rest of the text.

3. Lighter than Air

In the last section I promised to say a little more about aircraft which are lighter than air.

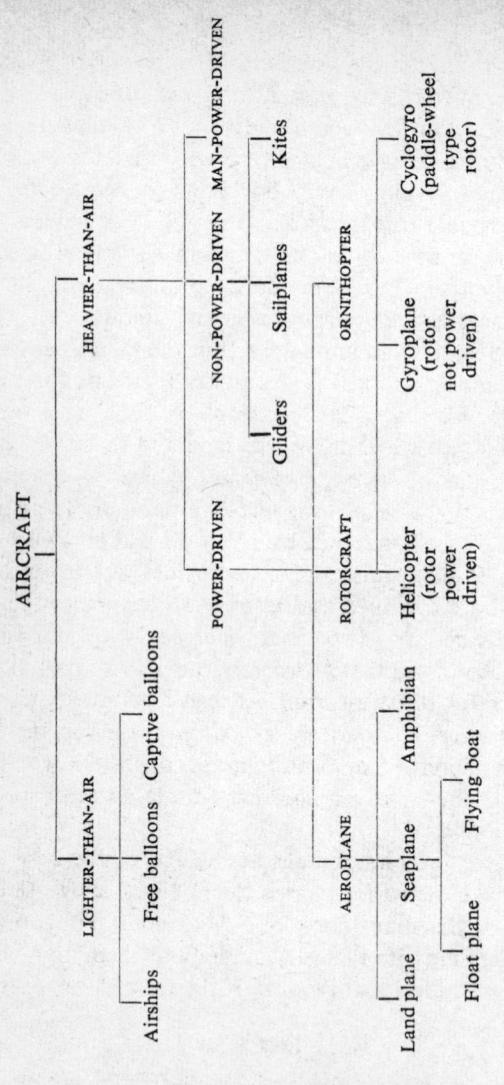

Fig. 1. Types of aircraft

LIGHTER THAN AIR

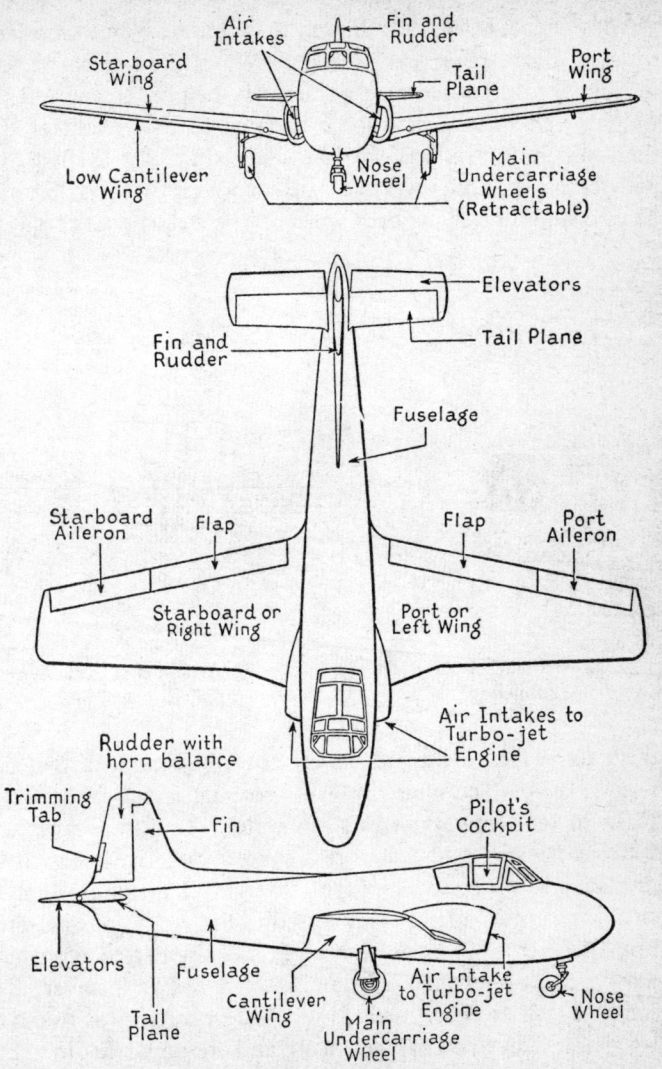

Fig. 2. **Parts of an aeroplane**

These depend for their lift on a well-known scientific fact usually called *Archimedes' principle*. When a body is immersed in a fluid, a force acts upwards upon it, helping to support its weight, and this upward force is equal to the weight of the fluid which is displaced by the body (Fig. 3). A fluid, of course, may be either a liquid, such as water, or a gas, like air. Thus a ship (or a flying boat when on the water) floats because

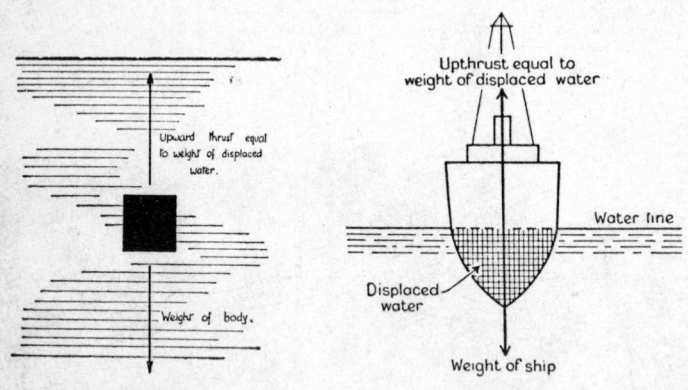

Fig. 3. **Principle of Archimedes**

Fig. 4. **Archimedes' principle applied to a ship**

the water which it displaces is equal to the weight of the ship itself (Fig. 4). The same ship will float higher out of the water when in sea water than in fresh water. This is because sea water is heavier, and therefore a smaller quantity needs to be displaced in order to support the weight of the ship. Only a small portion of a ship is immersed in the water, yet the same principle is true of bodies which are totally immersed and which may even be incapable of floating at all. For instance, if a lump of lead or other metal is weighed in water, it is found to weigh less than when weighed in air, and this apparent difference in weight is exactly equal to the weight of water which is

LIGHTER THAN AIR

displaced by the metal, thus proving that there is an upward thrust equal to the weight of displaced water.

An airship (Plate 35), balloon (Plate 34), or kite balloon (Plate 33) obtains its lift in precisely the same way. The envelope of the airship displaces the air, and therefore there is an upward force on the airship which is equal to the weight of the displaced air (Fig. 5). If this upward force is equal to the

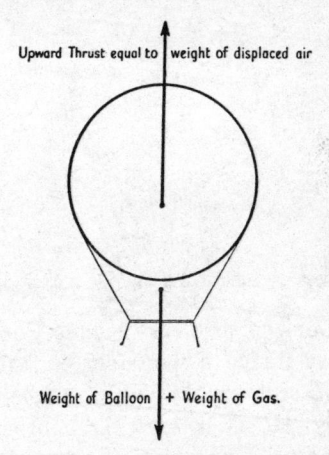

Fig. 5. **Archimedes' principle applied to a balloon**

weight of the airship, it will float; if the upward force is greater than the weight, the airship will rise; if it is less, it will fall. A cubic foot of air weighs only about 0·08 lb (roughly $1\frac{1}{3}$ oz), and therefore that is the greatest weight which one cubic foot can support. So you will soon see why it is necessary for the envelope of an airship to be so large and why the weight must be kept as small as possible. The R 100 and R 101, the last two airships to be built in Great Britain, had each a capacity of over five million cubic feet.

In order to keep the weight of the airship itself as small as possible it must in the first place be made of the lightest materials available, provided of course they are of sufficient strength. Secondly, a very light gas must be used in the envelope. Theoretically, the best thing which could be used in the envelope would be nothing, i.e. a vacuum; but in practice this cannot be done, because the pressure of the air outside

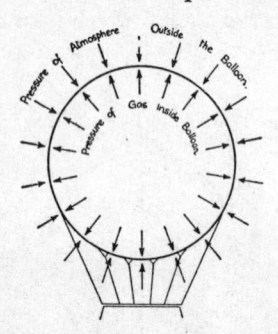

Fig. 6. **Pressure inside and outside a balloon or airship**

the envelope would be so great that the sides would cave in unless the skin of the envelope could be made tremendously strong, in which case it would weigh so much that no advantage would be gained. However, even the lightest gases can exert a pressure from the inside which will balance the pressure of the atmosphere from the outside (Fig. 6), and this means that the skin of the envelope need have very little strength, and therefore very little weight, provided it is gas-proof to prevent leakage in or out. The lightest gas in commercial use is hydrogen, and, for many years, this gas was always used in airships and balloons. Unfortunately, however, hydrogen is very inflammable, and its use added considerably to the dangers of lighter-than-air flying. So the gas helium came to be used, in spite of the fact that it is much more expensive and twice as heavy as hydrogen.

Hydrogen weighs about 0·0055 lb/cu ft and helium about 0·011 lb/cu ft, and in each case, of course, the weight of the

LIGHTER THAN AIR

gas tends to subtract from the lifting power of the displaced air. Thus, if an airship is filled with hydrogen, each cubic foot of envelope will support 0·0800 lb less 0·0055 lb, i.e. 0·0745 lb; but if filled with helium a cubic foot will only support 0·0800 lb less 0·0110 lb, or 0·0690 lb. If we multiply each of these by 5,000,000, they represent about 166 tons and 154 tons respectively. Thus the use of helium instead of hydrogen in an airship of this capacity will mean a loss of net lift of as much as 12 tons, and when it is remembered that the structure and engines of the airship itself will weigh over 100 tons, it will soon be realized that this loss of 12 tons is a very considerable proportion of the *useful* lift of the airship. However, so great was the fear of fire in airships, that the extra safety provided was held to justify the use of helium in spite of this consequent loss of lift.

We have said that a cubic foot of air weighs about 0·08 lb. Now, this is only true of the air near the earth's surface. As we ascend, the air becomes very much thinner and therefore a cubic foot will weigh less, and each cubic foot will consequently support less. So, if an airship is just able to float near the earth's surface, it will be unable to do so at a greater altitude, because the weight of displaced air will not be sufficient to support it. It is for this reason that ballast is carried; this can be thrown overboard to lighten the ship when it is required to climb. This is all very well while the climb is in progress, but what is to happen when we wish to descend? There is no means of taking on board extra weight, and therefore the only thing to do is to release some of the gas and allow air to take its place, thus decreasing the weight of air displaced, reducing the lift and allowing the ship to sink. It will be obvious that these processes cannot go on indefinitely, as neither the ballast nor the gas can be replaced until the airship returns to its base.

Another problem is that, owing to changes in the pressure

of the air outside the balloon or airship, it is not easy to equalize the pressures inside and outside the envelope at all heights unless the volume of the envelope can change. Thus it is that a toy balloon, filled with hydrogen at a reasonable pressure at ground level, expands as it rises and eventually

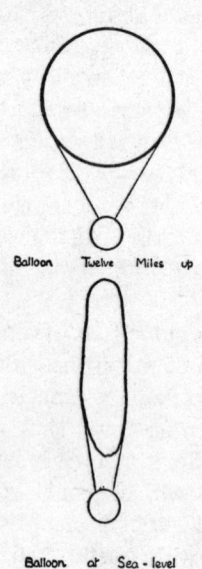

Balloon Twelve Miles up

Balloon at Sea-level

Fig. 7. **Stratosphere balloon**

bursts. To prevent such an occurrence with a real balloon it is only partially filled at ground level and presents the appearance shown in Fig. 7.

4. Lighter than Air—More Problems

These are some of the problems of lighter-than-air flight, but they are by no means the only ones. In order that an airship may carry a reasonable proportion of useful load it must be very large; the large ship means expense, difficulties of housing and manipulation on the ground, large head resistance, and very considerable structural design problems. All these difficulties, together with that of the fire risk, were courageously tackled in various countries, but repeated failure caused such losses in men and material in the period between the wars that in Great Britain, at any rate, we felt compelled to stop any further experiments on this type of aircraft. The wisdom of this policy was much disputed, but the fact remains.

Until the outbreak of the Second World War, experimental work on airships was still being carried out in Germany and the United States; in the latter country the metal-clad airship had been proved to be a practical proposition.

The war itself retarded rather than advanced experimental work on the subject, and the steady improvement which has taken place in aircraft of the heavier-than-air type is certainly likely to decrease the chances of a revival of interest in airships. But one can never be sure—as recently as 1958 a new non-rigid airship of about one and a half million cubic feet capacity was launched in the United States, and the Germans have never completely lost their faith in this means of transport.

Of the other lighter-than-air types the free balloon (Plate 34) may now be considered as obsolete except for scientific purposes such as the exploration of the highest regions of the atmosphere. There are also a few enthusiasts who still take part in ballooning as a sport.

The captive or kite balloon (Plate 33) was extensively used during the 1914–18 war as a means of observation for gunfire. After that war its chief use seemed to be to provide spectators at the Royal Air Force Displays with the never-failing attraction of seeing it brought down in flames. In the Second World War the captive balloon again played its part; this time as a means of protecting important towns and ships at sea from attacks by enemy aircraft; or, rather, to force raiding aircraft up to such a height that accurate bombing was rendered difficult. And although such balloons can have only a very limited use, either now or in the future, they still exist in reasonable numbers—which is more than can be said for the free balloon or airship.

5. The Atmosphere

But we cannot get much farther in understanding the problems of flight without considering in more detail the properties of the atmosphere on which it depends. The atmosphere is that very small portion of the universe which surrounds the surface of the earth with a belt of air—and it is only in this atmosphere that flying, as we have defined it in Section 2, is possible. The internal-combustion engine, whether piston or turbine, needs air in order to obtain its power; the lift of the aircraft, whether

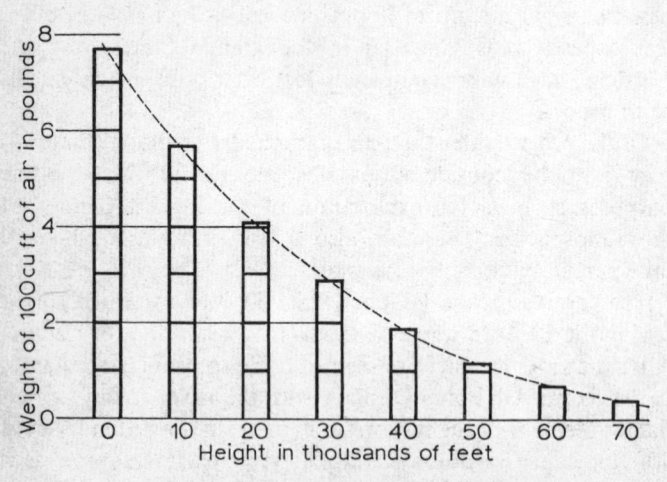

Fig. 8. **How density falls with height**

THE ATMOSPHERE   13

lighter or heavier than air, the controls, the stability, all depend on the air and the forces which it produces.

The most important property of the atmosphere, so far as flying is concerned, is its *density*. The way in which this falls off with height (Fig. 8) has already been mentioned in connection with lighter-than-air flight, but it is just as important

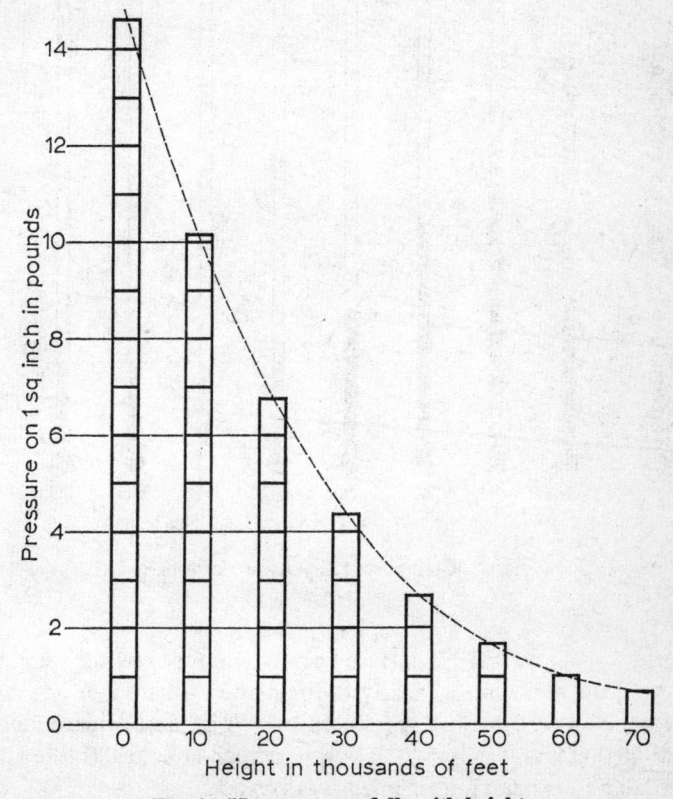

Fig. 9. **How pressure falls with height**

Although the curves of Figs. 8 and 9 look similar, they are not exactly the same: pressure falls off more rapidly than density

for heavier-than-air flight, and is clearly shown in the diagram; notice that whereas 100 cu ft of air weigh 8 lb at sea level, they weigh only 4 lb at 20,000 ft and less than $\frac{3}{4}$ lb at 60,000 ft.

Notice how the *pressure* also decreases with height (Fig. 9) —in fact, this is really the cause of the decrease in density,

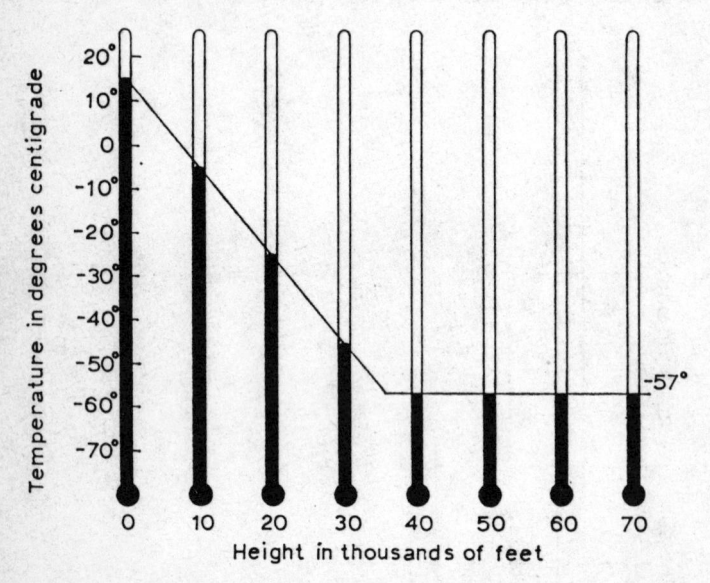

Fig. 10. **Change of temperature with height**

the air near the earth's surface being compressed by the weight of all the air above it, nearly 15 lb on every square inch at sea-level. As the pressure is released to 7 lb on each square inch at 20,000 ft and only 1 lb on a square inch at 60,000 ft, the air is able to expand and the density decreases.

The *temperature* also falls off, but in a rather curious way (Fig. 10). Up to about 36,000 ft above the earth's surface the

THE ATMOSPHERE

fall is quite regular (about 2°C, or 3°F, per thousand feet), then the fall suddenly ceases, and for greater heights the temperature remains fairly constant at about −57°C. At that temperature, however, there is not much consolation in knowing that it will not get any colder. This sudden check in the fall of temperature has resulted in the lower part of the atmosphere (that part with which we are most concerned in this book) being divided into two layers (Fig. 11), the one nearer the earth, in which the temperature is falling, being called the *troposphere*, the higher one, in which the temperature is constant, the *stratosphere*. The surface dividing the two is called the *tropopause*.

But perhaps the most aggravating feature of the atmosphere is its changeability—it is never the same from day to day, from hour to hour. For this reason we have been forced to adopt an average set of conditions (as shown in Fig. 11) called the *International Standard Atmosphere*. Although there may never be a day when the conditions of the atmosphere all the way up are exactly the same as those average conditions, they do serve as a standard for comparing the performances of aircraft. For instance, when a height record is attempted, the height allowed is not the height actually achieved but the height which, according to calculation, *would have been achieved if the conditions had been those of the International Standard Atmosphere*. So it is no good choosing a lucky day!

It is not easy to say how far the atmosphere actually extends, for the simple reason that the change from atmosphere to space is so gradual that it is impossible to decide on a definite dividing line; for this reason it is hardly surprising to find that estimates of the maximum height vary from 50 to 250 miles or more—rather a wide range. So far as aircraft are concerned, the higher we get, the more difficult does it become to go any higher. At record-breaking heights we already have to pump air into the engine, enclose the pilot in an air-tight suit, supply

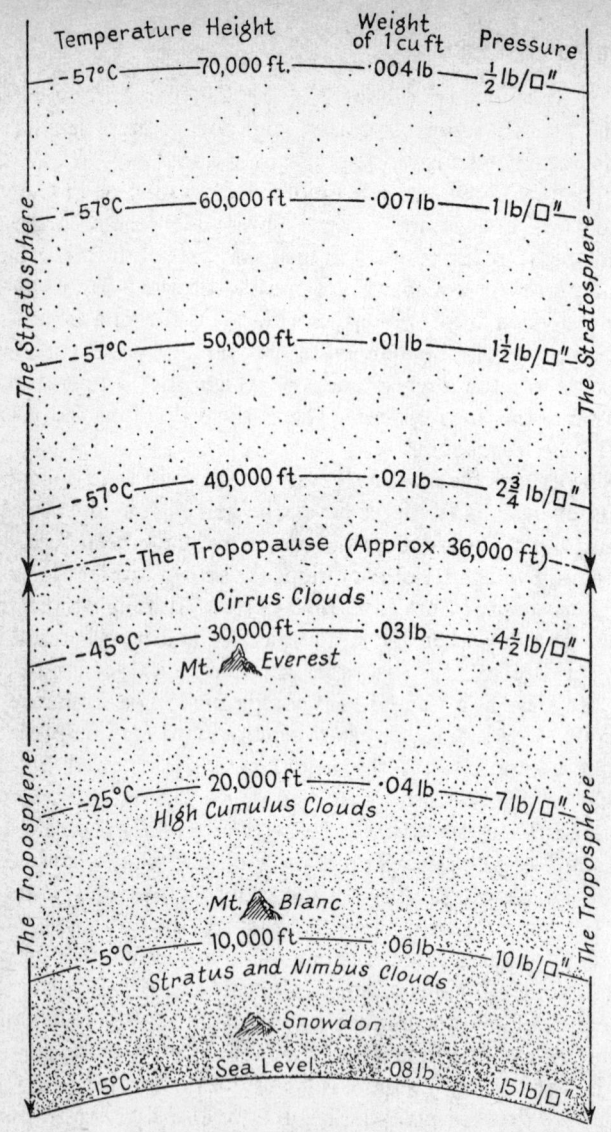

Fig. 11. **The International Standard Atmosphere**
(See footnote opposite)

him with oxygen, and heat his clothing artificially, while the aircraft itself can hardly get sufficient support in air that has not got one-quarter the thickness of the air near the ground.

Nor is it surprising that estimates of temperatures in even higher regions of the atmosphere vary very considerably— between temperatures both above and below anything known on earth—when the air is so thin it isn't the temperature of the air that matters so much as the temperatures of the outer surfaces of the aircraft.

But in these days of *missiles*, *satellites*, and *spaceships*, we have become very interested, not only in the upper reaches of the atmosphere, but in the space beyond it. These may not be aircraft (as we have defined the term), and although they may not even fly (according to our definition), no book on flight, with or without formulae, can any longer leave them out of consideration; we shall have more to say about them towards the end of the book.

6. Lift and Drag

But, for the present, let us return to earth and turn our attention to real aircraft, and more particularly to the aeroplane in its various forms.

In order that an aeroplane may fly, we must provide it with a lifting force at least equal to its weight. In that respect there is no difference between the aeroplane and the airship; it is in the method by which the lift is provided that the difference lies.

Take a piece of stiff cardboard (Fig. 12) and push it through the air in such a way that it is inclined at a small angle to the direction in which you push it, the front (or *leading edge*) being slightly above the rear (or *trailing edge*). You will find that

---

The figures given in Fig. 11 are only approximate, but they are sufficiently accurate to give a good idea of the changes in the atmosphere with height.

the result of pushing the cardboard through the air is to produce on it a force which tries to push it upwards and backwards. The upward part of this force we call *lift*, the backward part we call *drag* (Fig. 13).

It is quite likely that the upward force will be sufficient to lift the cardboard, which will thus be supported in the air. That

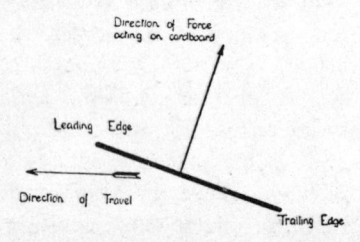

Fig. 12. **Principles of heavier-than-air flight**

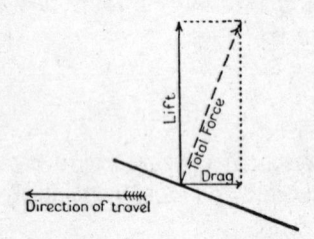

Fig. 13. **Meaning of lift and drag**

is how an aeroplane flies. So simple, isn't it? Yes—the cardboard is, in fact, acting just like the wings of an aeroplane.

What will happen if we release the cardboard? Try it for yourself, and you will soon see. It may continue its flight for a short distance—in fact, it may actually rise as it leaves your hand—but very soon it will cease to move forward, it will probably turn over, its leading edge going over the top, and then flutter to the ground. This shows that in order to obtain lift we must constantly push the cardboard forward, and in the real aeroplane this is provided for by the thrust. How thrust is obtained is explained in a companion book in this series, *Thrust for Flight*.

7. Air Speed and Ground Speed

In the last section we suggested that you should push the cardboard through the air. If you happen to try this simple experiment out of doors and if a wind is blowing, it will only be necessary to hold the cardboard *still* in a similar position, i.e. with its higher edge facing the wind. You will again feel the upward force, or lift, and the backward force, or drag, and if you release the cardboard it will behave very much as before. This is because it really amounts to the same thing whether the cardboard is pushed through the still air, or whether a stream of moving air moves past the cardboard. *The speed at which a body moves through the air, or at which the air moves past a body, is called the air speed. The speed at which a body moves over the ground is called the ground speed.* In our first experiment there was both a ground speed and an air speed, but in the second experiment there was an air speed but no ground speed, because the cardboard was held still relative to the ground.

We are so accustomed to thinking of speed and directions of movement *in relation to the ground* that it is very easy to forget that flying takes place in the air, and it is only movement *relative to the air* which matters when we are studying the flight of an aeroplane. I say "when we are studying the flight of an aeroplane," and you must understand clearly that this means when we are studying the principles and methods of flight; it is fairly obvious that if we wish to fly from London to Moscow it will make a considerable difference to the time taken whether the wind is with us or against us. In other words, the ground speed will matter very much when reckoning the time taken to fly between the two capitals, but the air speed, and therefore the lift and the drag, will be the same in both instances. An aeroplane is always travelling *against* a head wind. Thinking of it from a position on the ground, we may say that there is a following wind or a side wind, we may say that an aeroplane is flying "up wind" or "down wind"; but to the airman there is only a head wind. Anyone who has had experience of flying just above the clouds will have had convincing proof of this; he will have noticed how the clouds always seem to come to the aeroplane from the front, even though there may be a side wind or a following wind.

Once you understand it, all this will sound very simple and obvious, but I have emphasized it because I have found that many do not see daylight until the point has been pressed home.

Now ask yourself the following questions:

(a) The normal air speed of a certain aeroplane is 80 m.p.h. If it is travelling from west to east with a 100 m.p.h. westerly gale blowing behind it, in what direction will a flag on the aeroplane fly?

(b) In what direction will the flag fly if the gale is from the north and the aeroplane is still heading towards the east? (It will, or course, travel crabwise over the earth's surface.)

(c) In what direction will a flag fly in a free balloon which is flying in a steady wind of 30 m.p.h. from the north?

(d) An aeroplane has enough fuel to fly for 4 hours at 100 m.p.h. If there is no wind, how far can it fly out from base and get home again; that is to say what is its radius of action?

(e) Will the aeroplane of question (d) have the same radius of action if there is a steady wind of 20 m.p.h.?

(f) You are asked to handicap aeroplanes of different speeds for a race in which they will be required to fly

WIND TUNNELS

the theory of flight has advanced so much, the greatest designers hesitate to use any new device until it has been tried out on a model.

The most common method of experiment is to use a wind tunnel (Fig. 17), in which the model is supported while the air flows past it, the air being sucked through the tunnel by a fan driven by an electric motor. As we have already noticed,

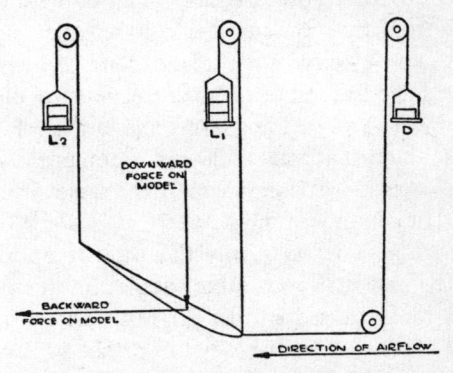

Fig. 18. **Principle of a wind-tunnel balance**

The weights L_1 and L_2 together measure the total downward force or lift; the weight D measures the backward force, or drag.

it is the relative air velocity which matters, so that for most purposes the air flowing past the stationary model will produce the same results as the model moving through the air. The forces on the model are measured by means of a balance, similar to an ordinary weighing machine, to which the model is attached by fine wires or thin rods (Fig. 18).

The results of wind-tunnel experiments are apt to be misleading for various reasons, the chief one being what is known as *scale effect*. The object of making experiments on models is to forecast the forces on the full-scale aeroplane when in the air. In order to do this we must know the laws which connect the forces on the model with those experienced in flight. It is fairly easy to form theoretical laws, and these, which will be mentioned in later paragraphs, are confirmed by experiment so long as there is not *much* difference between the size of the model and the full-scale aeroplane, or between the velocity of the wind-tunnel test and the velocity of actual flight. When the differences are great, and they often are, the laws seem to break down and our forecasts are found to be untrue. This is what is meant by *scale effect*, and it becomes more serious as the size and velocity of aeroplanes tend to increase. Fortunately we have learned to make corrections to allow for this error, and we are also building larger and larger wind tunnels— so large in fact that real full-size aeroplanes will go in them— but even so we cannot achieve the same air velocity in a tunnel as that of modern flight.

The reader may wonder why the wing is upside-down in Fig. 18. The explanation is quite simple; in this position the downward force caused by the air flow merely adds to the downward force due to the weight so that we only have to measure downward forces. If the wing were the right way up the lift due to the air flow would be upwards and the weight downwards and so we might have to measure forces in both directions.

In connection with scale effect you will hear highbrow people talking about *Reynolds numbers*. This is one of the instances where they try to pretend that they are talking about something which is far beyond your understanding. Don't believe it! A high value of the Reynolds number of a certain test is only a fancy way of saying that either the speed or scale of the test approaches full-scale value; the greater the speed, the greater the scale, the higher is the Reynolds number. Owing to the units used in calculating this number the numerical values are high, ranging from 100,000 or so in a test at low speed in a small wind tunnel to 20,000,000 or more for a large machine in high-speed flight. The term is an old one, dating back to Professor Osborne Reynolds, the famous British physicist of the nineteenth century, who discovered that the flow in water pipes always changed in character when the velocity multiplied by the diameter of the pipe reached a certain value—which came to be called the Reynolds number after him. The highbrows will say that I haven't told you the whole story. Nor have I—but I have told you enough to give you a good idea of what it is all about.

Another difficulty with wind-tunnel experiments is that all the details of a full-scale aircraft cannot be reproduced accurately on the model. Any reader who has made models will understand this difficulty. On an aeroplane there are many small parts, not to mention the roughness of the surfaces, and it is often these very details which are so important. There can be no way out of this difficulty except to make the models as large and as accurate as possible.

This leads us to yet another error. Both the difficulties already mentioned seem to suggest that we should make our models large, but, unfortunately, if the model is large, the tunnel must be *much larger still*, since otherwise the air is forced by the walls of the tunnel to flow quite differently from its flow in the free atmosphere. So once again we need large tunnels, and we are only limited by the expense involved and the power necessary to get high air velocity.

Some of our troubles can be overcome by working in compressed air, and there are compressed-air tunnels which can be pumped up to pressures of as much as 25 times that of the atmosphere. This is really an artificial means of increasing the Reynolds number while still keeping speed and scale within reasonable limits. In short, it helps to complete the story which we left unfinished earlier in this section, the truth being that the density (and viscosity) of the fluid also affects the Reynolds number of the test.

One would naturally expect the most valuable experiments to be those made on full-scale aeroplanes in flight. While it is clear that this must be the eventual test, there is much to be said against it—when compared with the wind tunnel—for experimental purposes. Flying with new and untried devices may be dangerous, and it will certainly be expensive. The air is never steady, nor are conditions the same from day to day, and one cannot test separate parts, such as a wing, a strut or a wheel. So, with all its faults, there is something to be said for the wind tunnel after all.

10. Smoke Tunnels

One of our difficulties in experimental work is that *we cannot see the air*, and it is the way in which the air flows that is so important (Fig. 19). If the air were visible, there is no doubt

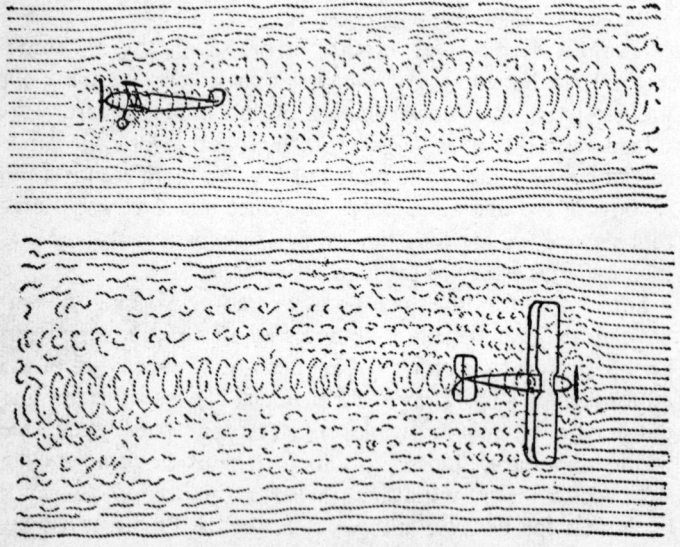

Fig. 19. **Seeing the air**

that many so-called aeronautical discoveries would have been obvious to everyone. For this reason attempts have been made to show the flow of air by introducing jets of smoke, and this is best done by using a small *smoke tunnel* and projecting the results by means of a lantern on to a screen. Very effective demonstrations can be made in this way, but the difficulty is to find a suitable smoke. Most smoke that can be produced in large quantities, and is about the same density as air, is objectionable in some other way. After experiments with complicated and difficult chemicals, the most satisfactory results were eventually obtained with smoke produced by heating paraffin or burning cardboard or rotten wood.

Although the reader may have no chance of seeing experiments in a smoke tunnel, he should always watch dust or leaves being blown about, or tobacco smoke; a lot can be learnt in this way. Very often short streamers, or tufts of wool, are attached to models or aeroplanes, as all these are useful aids towards "seeing the air."

11. Air and Water

Sometimes experiments are done by moving models through water, because, strange as it may seem, water behaves very much like air except that the velocity need not be so high, and this is an advantage from the experimental point of view. To get similar results a body need only move through water at about one-thirteenth of the speed at which it moves through air.

Quite apart from the scientific use of water as a means of aeronautical experiment, it is much more suitable than air for amateur observation. Move your hand through air and nothing *appears* to happen—in fact, *quite a lot does*—but move your hand through water and you can not only see the effect but you can feel the resistance to motion. There is no need to give to Archimedes all the credit for making discoveries in his bath; you can do the same yourself and not only can you discover, or rediscover, his principle (on which, as we have said, lighter-than-air flight depends), but you can discover, too, many of the principles of heavier-than-air flight as outlined in this book.

The reader may wonder at the idea of water behaving like air; if he does so, he certainly deserves a word of explanation. Both water and air are fluids; but water is a liquid and air is a gas, and one of the differences between a liquid and a gas is that the former is, for all practical purposes, incompressible, whereas the latter is easily compressed. Is not this question of compressibility important in flight? The answer is, at low speeds, *no;* at high speeds, *yes.* That is all very well, but high speed and low speed are relative terms; where is the dividing line? You may be surprised at the answer. *Low-speed flight*—that is to say, flight in which the compressibility of the air is not of practical importance—*is flight at speeds less than that at which sound travels in air.* *High-speed flight*—that is to say, flight in which the compressibility of the air is of importance—*is flight at speeds greater than that at which sound travels in air.* What is this speed? And why is it so significant? The first of these questions is easy to answer—about 760 m.p.h. or 1,100 ft/sec—not exactly dawdling! The second question needs and deserves a longer answer, and it will be given in some of the later sections. Suffice it now to say that sound, which is in effect a compression of the air, travels or is transmitted through the air on a kind of wave which compresses first one part of the air, then the next, and so on. When a body moves through the air at speeds lower than the speed of sound these sound or pressure waves go out in front and warn the air that the body is coming; the air then simply gets out of the way, passing on one side of the body or the other, just as water divides when a ship passes through it. The air is not compressed, and behaves just as if it were incompressible—like water. But when a body travels at speeds above that of sound, the warning wave does not travel fast enough to get ahead of the body, so the air, instead of dividing and passing smoothly past the body, comes up against it with a shock and is compressed.

Now, most aeroplanes even in these days cannot fly at the speed of sound, and even those that can must start and end their flight below that speed—let us hope that they always will!—and so the subject that we are still most concerned with is that of low-speed flight, that is flight in which the air behaves as if it were incompressible and in which we can therefore learn from experiments in water. Most of this book is devoted to this kind of flight, but the time is past when an author can avoid the obligation of saying something about the other kind, and this obligation will be fulfilled to the best of my ability in later sections.

While discussing the subject of air and water it may be appropriate to mention a type of vehicle which is actually supported by wings under water—the *hydrofoil craft* (Plate 49). We can hardly call this an aircraft, but if we substitute "driven through the water" for "driven through the air", it fulfils the definition of an aeroplane as given on page 2. The similarities —and differences—between hydrofoil craft and aircraft are so interesting that a book on *Hydrofoils* has been included in this series.

12. Centre of Pressure

After this long but important diversion, let us return to our cardboard. If, as we push it through the air at a small angle— this angle, by the way, is called the *angle of attack* or *angle of incidence* (Fig. 20)—we hold it at the centre of each end, then not only shall we feel an upwards and backwards force exerted upon it, but it will tend to *rotate*, its leading edge going over the top. Similarly, if we try to make it glide of its own accord, it will turn over and over. This is because the effective or resultant force acting upon it is *in front of the centre-line*, whereas we are holding it on its centre-line, or, when it is left free to fly by itself, its weight is acting downwards at the centre. If we hold it farther forward, or if we add weights to it so that its centre of gravity is farther forward, we shall eventually

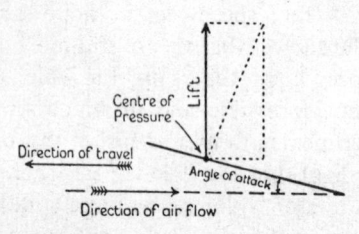

Fig. 20. **Angle of attack and centre of pressure**

find that it tends to turn the other way, the nose dipping downwards. With a little practice we can find a position such that it does not tend to turn either way, and then we have found what is called the *centre of pressure* (Fig. 20).

13. Stability and Instability

When the centre of pressure and the centre of gravity coincide, the plane is balanced, or is *in equilibrium* (Fig. 21*a*). If the centre of pressure is in front of the centre of gravity, it is said to be *tail-heavy* (Fig. 21*b*); whereas if the centre of pressure is behind the centre of gravity, it is *nose-heavy*. At present we are talking about a piece of cardboard; we are doing things in a simple way at first, but we are all the time learning big principles, and what is true of the cardboard is equally true of an aeroplane weighing many tons.

Now, all would be very simple if the centre of pressure always stayed in the same place, but unfortunately it does not.

STABILITY AND INSTABILITY

As we alter the angle of attack, i.e. the angle at which the plane strikes the air, *the centre of pressure tends to move*. We shall investigate the reason for this later; at present, let us be content with the fact that it *does*. If, as we increase the angle, the centre of pressure moves forward, then it will be in front of the centre of gravity and will tend to push the nose farther

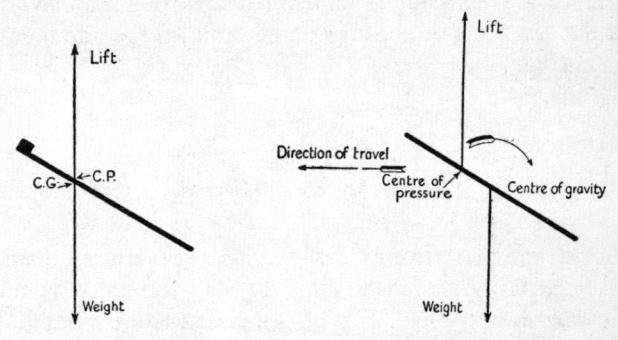

Fig. 21a. **Balance** *Fig. 21b.* **Tail-heavy**

upwards, thus increasing the angle still more. This in turn will cause the centre of pressure to move *farther* forward, and this —well, you can guess the rest. This is called an *unstable* state of affairs—the mere fact that things become bad makes them tend to become worse. If, on the other hand, as we increase the angle, the centre of pressure moves backwards, it will then be behind the centre of gravity and will tend to push the nose down again and restore the original angle. This is called a *stable* state—when things become bad, influences are set up which tend to make them become better again. As before, what is true of the cardboard is true of the aeroplane—if we want the aeroplane to be stable, and you can probably guess that we do, then we must arrange for the latter conditions to apply. How? That is a long story; but it will all come out in due time.

14. The Wing Section

Everyone nowadays knows that, although we still call an aeroplane wing a "plane," it is not, in the geometrical sense a "plane" at all. It is a curved or *cambered* surface—in fact, it is really made up of two surfaces, each with a different curve or camber. The technical name for such a wing is an *aerofoil*, and the cross-section through an aerofoil is called an *aerofoil section* (Fig. 22).

Fig. 22. **An Aerofoil Section**

There are two reasons for curving the surface: first, *a curved surface gives much better lift*, and secondly, we must have *thickness* to give strength to the structure. Some old books on the subject devoted a lot of space to the study of the flat plate; and in the last edition of this book we were rash enough to say that no flat surface had ever been used or was ever likely to be used for real aeroplanes. What a wonderful example of how careful one has to be in this subject—instead of *real* aeroplanes we ought to have said *low-speed* aeroplanes, because in fact many supersonic aerofoil sections have some flat surface, though of course they must still have thickness to give them strength.

It is true that we began our study with a flat piece of cardboard, but it did serve to explain terms like angle of attack, lift, and drag. Besides, there was another reason: bend it into a curved wing section and try it for yourself. It won't fly so well as it did when it was flat—in fact, the chances are that it will turn over on its back. You then may try to readjust the weight because the centre of pressure is in a new position. If you do, it will probably turn over in the other direction. It

AIR FLOW OVER A WING SECTION 35

has become *unstable* (Fig. 23), whereas, as a flat plate, it was slightly *stable*. We have discovered the only disadvantage of the curved surface for aircraft that fly below the speed of

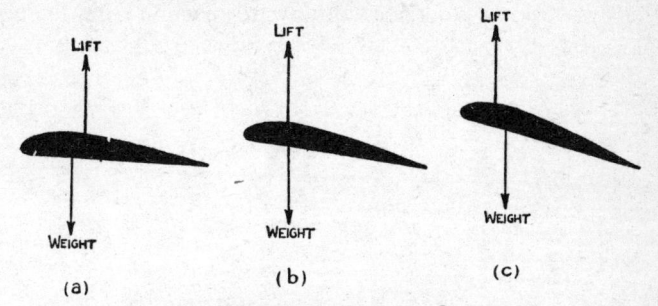

Fig. 23. **Movement of centre of pressure**
(*a*) Small angle—nose-heavy.
(*b*) Medium angle—balanced.
(*c*) Large angle—tail-heavy.

sound; but before we enlarge on that, let us turn to a further investigation of its *advantages*.

15. Air Flow over a Wing Section

If we wish to go *upwards* we must push something, or try to push something, *downwards*. In climbing a rope one gets a hold of it and pulls oneself upwards by trying to pull the rope downwards. In going up a flight of stairs one puts one's foot on to the next stair and attempts to push it downwards, and the stair exerts an upward reaction by which one is lifted. It is true that in these instances neither the rope nor the stairs actually move downwards and it is better that they should not do so; but there are instances, such as in ascending a sandy slope, where for each step upwards sand is pushed downwards. A drowning man will clutch at a straw—it is his last dying effort to get hold of something and pull it downwards so that he can keep himself up.

An aeroplane is no exception to these rules; the wing is so designed, and so inclined, that (in passing through the air) it will first attract the air upwards and then push it downwards and by so doing experience an upward reaction from the air.

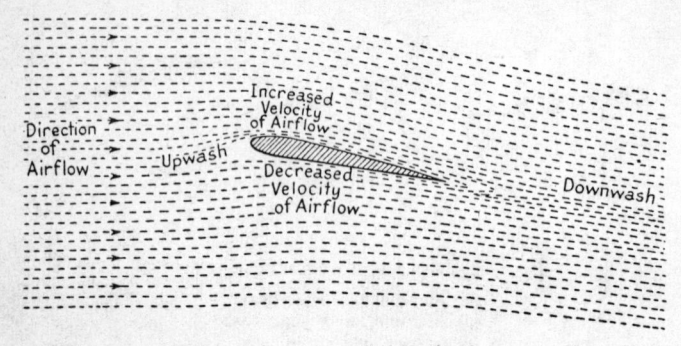

Fig. 24. **Air flow over an aerofoil inclined at a small angle**

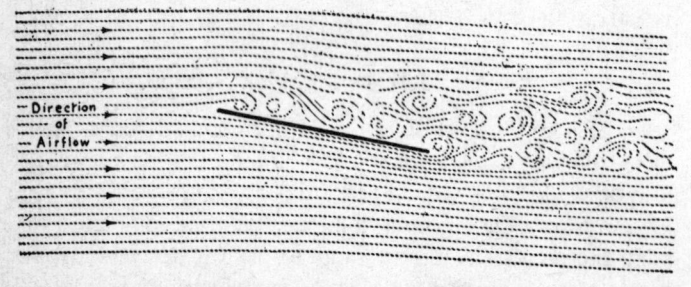

Fig. 25. **Air flow over a flat plate**

It is a simple example of one of *Newton's laws*—"To every action, there is an equal and opposite reaction." The downward flow of air which leaves an aeroplane wing is called *downwash*.

Now, the greater the amount of air which is deflected downwards by a wing in a given time, the greater will be the upward reaction, or lift; on the other hand, the greater the disturbance caused by the motion of the wing through the air, the greater will be the resistance to motion, or drag. Therefore the aim in the design of a wing, and in choosing the angle, is to secure as much downwash as possible *without at the same time causing eddies or disturbance*. This is where the curved aerofoil is superior to the flat surface (Figs. 24 and 25), and this also explains why the angle of attack used in flight is so small. The gradual curvature of the wing section entices the air in a downward direction and prevents it from suddenly breaking away from the surface and forming eddies, and although a larger angle would give more lift, it would create more disturbance and cause more drag.

16. Pressure Distribution round a Wing Section

Notice how effective the *top* surface of the wing is in curving the air flow downwards; the bottom surface acts in much the same way on both the aerofoil shape and on the flat plate, but it is the top surface of the aerofoil which scores.

We have, so far, considered the reaction on the wing as if it were a single force acting at a place called the centre of pressure,

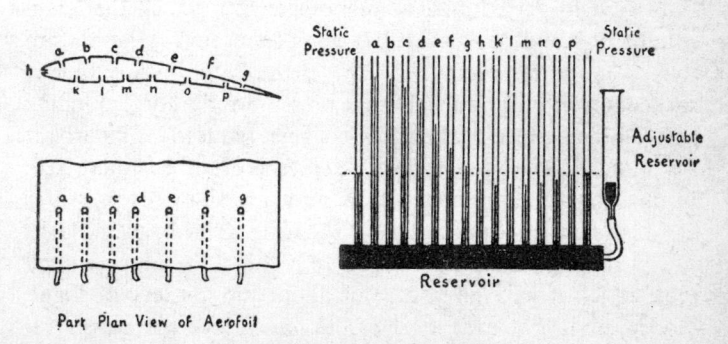

Fig. 26. **Pressure plotting**

but it is in reality the sum total of all the pressure acting upon the surface of the aerofoil, this pressure being distributed all over the surface. Distributed—yes; but *not*, by any means, *equally* distributed. This can easily be shown by what is known as *pressure plotting*. Small holes round the wing are

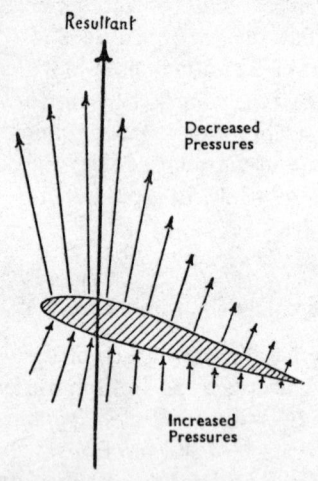

Fig. 27. **Distribution of pressure over a wing section**

connected to glass tubes, or manometers, in which there is a column of liquid, the glass tubes being connected at the bottom to a common reservoir (Fig. 26). If the liquid in any tube is sucked upwards, it means that the pressure at the corresponding hole on the surface of the wing has been reduced; similarly, if the liquid is forced downwards, the pressure has been increased. In this way a kind of map can be drawn to show the pressures at different parts of the surface of the wing. The diagram shows such a map for a typical wing section (Fig. 27). Have a good look at it; it will tell you a lot about the mystery of flight. Notice, first, that over most of the *top* surface the *pressure is decreased*—this is due to the downward curvature of the air;

on the *bottom* surface, however, the air is pressed downwards, and there is an *increase of pressure*. Notice that the decrease in pressure on the top surface is much more marked than the increase underneath, and thus the *top surface contributes the largest proportion of the lift*. This is only another way of saying what we had already noticed, namely that it is the top surface which is chiefly responsible for the downwash. The diagram of pressure distribution also confirms another previous discovery: it is quite clear that the majority of the lift, both at top and bottom surfaces, comes from the front portion of the wing, and therefore when we replace all this distributed pressure by a single force, we must think of that force as acting in front of the centre of the aerofoil—in other words, the centre of pressure is well forward (notice how we showed this in the earlier diagrams). All this confirmation of what we had already discovered should give us confidence, and we need confidence in this subject because, although it is all founded on simple laws of mechanics, it is full of surprising results and unexpected happenings.

I do not want, in this book, to worry you with formulae, figures or mathematics of any kind. You will find all that in more advanced books on the subject. But there has been so much misconception as to the *values* of the pressures round an aerofoil that I would like to put your mind at rest on that point at any rate. The misconception arises chiefly owing to the habit of describing the decreased pressure on the top surface of an aerofoil as a "vacuum" or "partial vacuum". Now, a vacuum means the absence of all air pressure; a vacuum on the upper surface would cause water in the corresponding tube in the manometer to be "sucked up" to a height of about 36 ft. In actual practice, the column of water rises three or four inches, so it is not much of a vacuum, hardly even worthy of the name of "partial" vacuum. Another way of looking at it is that a vacuum on the top surface would result in an effective upward pressure—*from the top surface only*—of nearly 15 lb/sq in, whereas the actual lift from an aeroplane wing may not even be as much as 15 lb/sq ft—and there are 144 square inches in a square foot. No, the truth of the matter is that the pressures round an aeroplane wing are but small variations in the usual atmospheric pressure of about 15 lb/sq in, and that is why such a large wing surface is necessary to provide the lift required.

Perhaps we ought also to explain that where you see the arrows on the top surface pointing upwards you must not think that there is really a kind of upward negative pressure on this surface—it is impossible to have a pressure less than nothing. What these upward arrows mean is that the pressure is reduced *below the normal atmospheric pressure*, and this, *in effect*, is producing an upward pressure. The normal atmospheric pressure is about 14·7 lb/sq in, so that when the wing is *not* moving through the air there will be a *downward* pressure on the *top* surface of 14·7 lb on every square inch, and an *upward* pressure on the *bottom* surface of the same amount. These two cancel out, and the net effect of the pressures is nil. Now, when the aerofoil is pushed through the air, the pressure on the top surface is still downwards, but it is *less than* 14·7 lb/sq in, whereas the pressure on the bottom surface is still upwards but *more than* 14·7 lb/sq in, and so there is a net upward pressure equal to the difference between these two, and the arrows are intended to show that this upward pressure is contributed both by the decrease on top and by the increase underneath—more by the former than the latter.

17. The Venturi Tube

The reader may feel that he would like a little more explanation as to *why* the pressure is decreased above the aerofoil and increased below it. All we have said so far is that it is due to

THE VENTURI TUBE

the downward curvature of the air flow, and this is certainly one way of looking at it. But perhaps a better way is to compare it with similar examples of the same sort of thing. Do you know what a *venturi tube* is? In case you do not, here is a picture of one (Fig. 28). It is a tube which has an inlet portion, gradually narrowing, then a throat or neck, followed by the outlet, which gradually widens. In a well-designed venturi

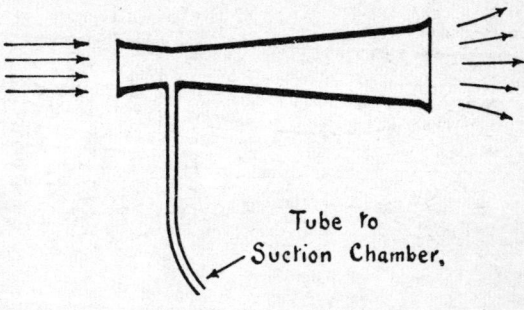

Fig. 28. A venturi tube

tube the outlet is usually longer than the inlet. The tube is so shaped—and it must be done very carefully—that air or other fluid which passes through it continues in steady streamline flow; if large eddies are formed, the whole idea of the tube breaks down. Now, it is quite clear that the same amount of air must pass through the throat as passes into the inlet and out of the outlet. Therefore, since the cross-sectional area of the tube at the throat is less than at inlet and outlet, it follows that one of two things must happen—either the fluid must be compressed as it passes through the throat, or it must speed up. The throat is after all what is commonly called a bottle-neck, and we all know from numerous examples in ordinary life the sort of things that can happen at a bottle-neck. Think, for instance, of a gate or narrow passage at the exit from a football ground. The badly disciplined crowd will try to push through the gate, they will be compressed, and, quite apart from the discomfort, the whole process of getting away from the ground will be delayed. The well-disciplined crowd, however—if there is such a thing—will move faster as they approach the gate, pass through it at a run, then slow down again as the path widens. Contrast, too, the way in which the traffic tries to push its way through some of the notorious bottle-necks in

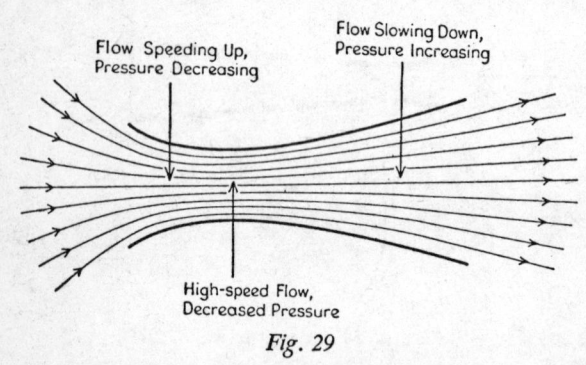

Fig. 29

Flow Speeding Up, Pressure Decreasing

Flow Slowing Down, Pressure Increasing

High-speed Flow, Decreased Pressure

the London streets, and the well-disciplined speed-up through the Mersey tunnel.

Now which of these two things happens when a fluid passes through the venturi tube? Is it compressed at the throat, or does it flow faster? The answer, in the case of water, is clearly that it flows faster; first, because water cannot be compressed (appreciably, at any rate); secondly, and perhaps more convincingly, because there are so many practical examples in which we can watch water and *see* how it speeds up as it passes through the throat—stand on a bridge and watch the water as it flows between the supporting pillars. The reader may not be so easily convinced about air, but the fact is that the patterns of air and water flow through a venturi tube are almost exactly the same (Fig. 29)—so much so that indistinguishable photographs can be taken—and measurements of

THE VENTURI TUBE

the speeds show that air speeds up just like water, and, as already explained, behaves as though it were incompressible—provided always that we are considering a speed of flow well below that at which sound travels.

And what is the result of this speed-up of the flow at the throat? Our pressure plotting experiment, now applied to the venturi tube, gives a convincing answer to that question, though it is not so easy to explain *why* it happens. At the throat the pressure which the air exerts on the sides of the tube is less than at outlet or inlet; in fact as *the velocity increases, the pressure on the walls of the tube decreases*, and vice versa. Why? The answer you will usually be given is simply *Bernoulli's theorem*. That doesn't sound very convincing; and what is Bernoulli's theorem? Well, you have probably heard of the idea of the conservation of energy—that energy may be transformed from one form into another but that the sum total of all energy in the universe remains the same. Some people will tell you that it isn't true; but don't worry about that, it is true enough for the purposes for which we are concerned with it. Well, Bernoulli's theorem is a kind of special application of this principle in so far as it concerns the flow of fluids—or rather the streamline flow of fluids, because if the flow is turbulent the theorem breaks down. In effect, the theorem states that, in streamline flow, the sum of the pressures exerted by the fluid remains constant. Now, a fluid can exert pressure for two reasons: first, because of its movement—this is the pressure that we feel when wind blows against our faces—secondly, because of the energy stored in it which makes it exert pressure on the sides of a vessel even when it is not moving—this is the pressure exerted on the envelope of a balloon, on the walls of a pneumatic tyre, or, to use the most common example, the ordinary atmospheric or barometric pressure. The pressure due to movement we will call *dynamic pressure*, the other the *static pressure*.

So, according to Bernoulli's theorem, the sum of the dynamic and static pressures remains constant—therefore, as the velocity (and the dynamic pressure) goes up, the static pressure must come down. We cannot prove the theorem here, but, what is perhaps more convincing, we can give several examples of its truth in practice. This is probably advisable, because it is one of those scientific principles which some people think are contrary to common sense—which seems to suggest that common sense is more common than sense, but that is by the way. Have you noticed how the dentist attaches a tube to an ordinary tap and in that tube is a small glass venturi, from the throat of which another tube leads to your mouth? The flow of water through the venturi causes a decrease in pressure which sucks moisture out of your mouth. Have you ever noticed how wind blowing through a narrow gap tends to suck in leaves and dust towards the gap? Have you seen a draught through a slightly open door close the door, rather than open it, as common sense might suggest? Have you noticed how in a whistle, or in most wind instruments, air is sucked in towards the throat in the instrument? Two ships passing close to each other tend to be sucked together, and this has often been the cause of collisions; similarly a ship passing close to a wharf tends to be sucked in towards the wharf.

But the best examples of all are from our own subject. Consider the wind tunnel, for instance. When the air is rushing through it, the pressure of the air outside is greater than the pressure at the narrowest part of the tunnel where the air is flowing fastest. If you doubt this, try to open a window or door in the tunnel, and you will soon know all about it. Venturi tubes themselves—sometimes double venturi tubes, a little one inside a big one—are used for all kinds of suction instruments, for measuring air speed by suction, for driving gyroscopes by suction. The choke tube in a carburettor is a perfect example of the practical use of a venturi tube. And last, the aerofoil which we are trying to explain. Here there is no obvious venturi, but by looking carefully at the way in which the air flows (Fig. 24) you will notice that the decreased pressures are where the streamlines are close together, where the air is flowing with higher velocity as at the narrower portions of the venturi. As a general rule, the air flows faster all over the top surface, and slower all over the lower surface. The greatest velocity of all is at the highest point of the camber on the *top* surface, and here is the least pressure, as at the throat of the venturi. But—let us emphasize this once again, because it is important—the best results will only be obtained if the streamlines are kept flowing close to the surface; as soon as they break away, on both aerofoil and venturi, there will be less decrease of pressure, less suction.

One of the best ways of thinking about air, or water, flowing through a venturi tube or over an aerofoil is to think of how the changes of pressure affect the flow rather than—as we have done so far—of how the flow affects the pressure. It is, after all, rather like the chicken and the egg—one doesn't know which came first. A fluid flows easily from high pressure to low pressure; there is, in technical terms, a favourable pressure gradient—it is flowing downhill so far as pressure is concerned. This is what is happening between the entrance to the venturi and the throat, or over the top surface of the aerofoil as far as the maximum camber—the air is free-wheeling, it likes it. But after the throat, or the point of maximum camber, the pressure is increasing, the pressure gradient is adverse, the air is trying to go uphill, if we are not careful it will stall—yes, just that!

18. Why the Centre of Pressure Moves

If we follow up this "pressure plotting" idea we shall find not only confirmation, but explanation, of another phenomenon that may have puzzled us. If we plot the pressure round the aerofoil at different angles of attack we shall find that the pressure distribution changes, and that it changes in such a way that as we increase the angle (up to a certain limit) the tendency is for the most effective pressures to move forward, thus causing the *resultant* forces to move forward, and so accounting for the instability of the aerofoil (Fig. 23). On the other hand, if we plot the pressure round a flat plate—not an easy thing to do—we find that the pressure distribution changes in a different manner, the resultant force tending to move backward as the angle increases, making the flat plate stable (*see* Section 14).

19. Stalling or Burbling

In Section 15 we mentioned that the angle of attack used in flight was a small one because "although a larger angle would give more lift, it would create more disturbance and cause more drag." The question of what is the best angle needs a little further investigation.

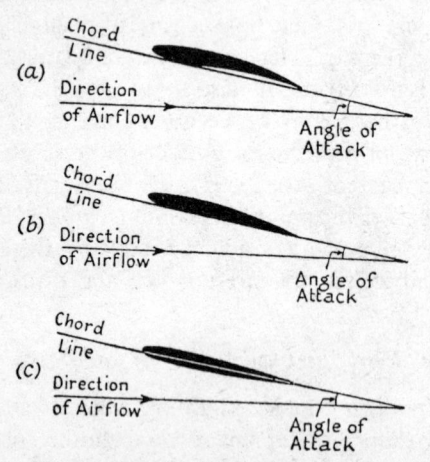

Fig. 30. **Chord line and angle of attack**

STALLING OR BURBLING

Before going into this we ought to mention that it is not so easy to define what we mean by "angle of attack" now that we have the curved aerofoil surfaces instead of our original flat plate. Clearly we must choose some straight line to represent the aerofoil—but what straight line? It sounds a simple

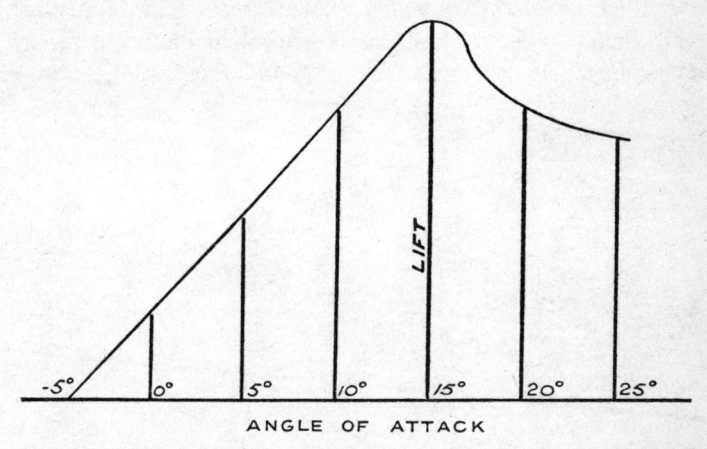

Fig. 31. **Variation of lift with angle of attack when the air speed remains constant**

question, but it has not been at all easy to solve, largely because methods which are satisfactory when considering the subject theoretically are quite impracticable to those whose duty it is to take actual measurements on the aeroplane. To cut short a long story, we can only say that different *chord lines* are used for different-shaped aerofoils (see Fig. 30), and the *angle of attack (for aerofoils) is defined as the angle which the chord line makes with the air flow.*

Now, if we increase this angle, what will happen?

Again it is not quite such an easy question as it sounds, and an enormous amount of experimental investigation has been made in order to answer it. So far as the lift is concerned, it increases as we increase the angle (provided that the air speed remains constant) *but only up to a certain limit*; after this it begins to fall off. Although the actual amount of lift given by the wing when this maximum limit is reached varies tremendously according to the shape of the aerofoil section, it is rather curious that most wings, whatever their shape of section and whatever the air speed, reach their maximum lift at about the same angle, usually between 15° and 20° (Fig. 31).

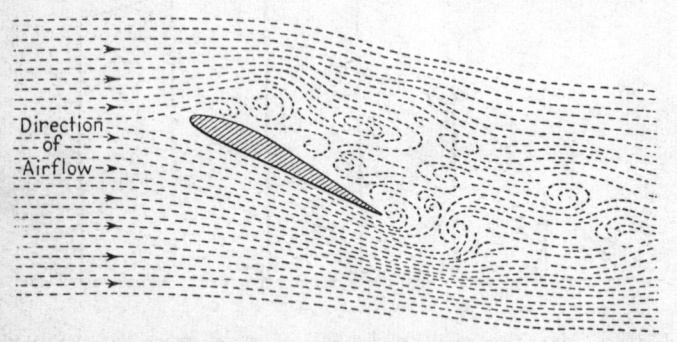

Fig. 32. **Burbling air flow over a wing inclined at a large angle**

Now, why does the lift fall off after this angle has been reached? One would think that the increasing angle would create more downwash and consequently more lift. It is rather natural that aeronautical engineers should have spent much time and study on this phenomenon, because flight would become very much easier and very much safer did it not occur. By watching the flow of air over the wing—using smoke or streamers so that they can see the type of flow—they have discovered that *when this critical angle is reached the flow over the top surface changes—quite suddenly—from a steady streamline flow to a violent eddying motion, with a result that much of the downwash, and consequently the lift, is lost* (Fig. 32). As one might expect, the drag, by the same token, suddenly increases.

Exactly the same thing happens in a venturi tube if we make the throat too narrow, or try to expand the tube too suddenly after the throat. In this connection it is interesting to note that, although the front part of the wing section, and the entry and throat of the venturi tube, seem to experience nearly all the effect so far as reduction of pressure is concerned, they are entirely dependent for this effect on the shape and angle of the rear portion of the wing and the expanding exit portion of the venturi. It is no good saying the front part gives the results, therefore why worry about the rear part; why not, even, cut it off? It is the front part that will suffer if you do.

The truth is that the flow is very sensitive to the exact shaping and angle and attitude of the whole system, whether it be a wing section or a venturi tube, and immediately we attempt to go too far it shows its objection by breaking down into turbulent flow—and so spoiling everything. If the hill is too steep, it just won't go up it!

This phenomenon is called *stalling* or, rather appropriately, *burbling*—it is one of the greatest problems of flight.

20. Lift and Drag again

Now, it is the air flow and the consequent pressures, as described in the preceding sections, that give us at one and the same time the *lift* which enables us to fly, in heavier-than-air craft, and the *drag* which tries to prevent us from doing so. Both are really part of the same force, but owing to their very different effects it is important to distinguish between them.

One of the unfortunate aspects of this subject, from the point of view of those who learn it or teach it, is that one constantly has to correct or modify one's original ideas. What I am going to tell you now is a glaring example of this. You will have gathered from what you have read that lift is an upward force and drag is a backward force. You will probably claim—not without justice—that I have told you so (see Section 6). Now I have got to tell you that that idea isn't true—or, rather, that it is only true in one particular case, i.e. when the aeroplane is travelling horizontally (even then the lift may be *downwards*, as it was on the model in Fig. 18). The real definition of lift is that it is *that part of the force on a wing (or an aeroplane or whatever it may be) which is at right angles to the direction of the air flow*—or, what comes to the same thing, at right angles to the direction in which the aeroplane is travelling. Similarly, *drag is that part of the force which is parallel to the direction of the air flow*. So you will see that the upwards idea of lift and the backwards idea of drag are only true for horizontal flight. In a nose-dive it is lift which will be horizontal and drag vertical. So far as lift is concerned, the correct definition is a rather silly one, because in ordinary language the word lift surely implies *upwards*; that is really my excuse for not telling you the truth earlier, because I did not want you to get the impression that it was a silly subject. Perhaps, by now, you have already realized that it is!

21. Effects of Speed

Both lift and drag increase with speed. Everyone knows this—at any rate so far as drag is concerned; one has only to try to pedal a cycle against winds of different velocities, and there can no longer be any doubt. In view of such common experience, it is rather surprising that most people seem to underestimate *how much* the resistance increases as the speed increases. They will usually tell you that if the speed is doubled they would think that the drag would be about doubled, perhaps a little more, perhaps a little less. This is very much of an underestimate, the truth being that for double the air

EFFECTS OF SIZE

speed the drag and the lift are about four times as much; for for three times the speed they are nine times and for ten times the speed they are multiplied by a hundred (Fig. 33).

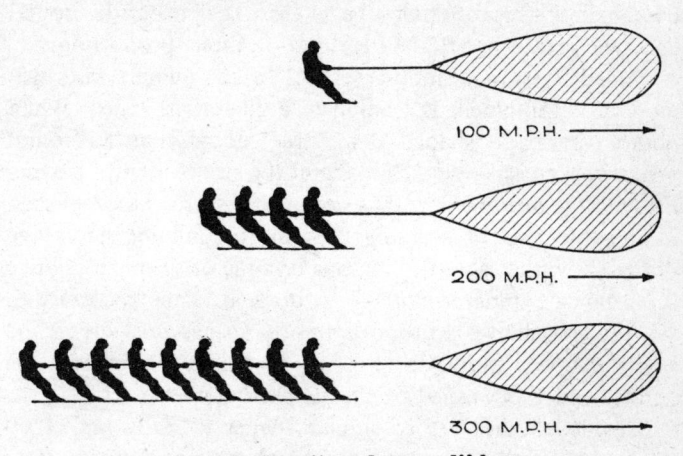

Fig. 33. **The "speed squared" law**

The men represent the resistances holding the bodies *back* at the various speeds; there must, of course, be corresponding forces pulling the bodies *forward*.

This is called the *speed squared law—the lift and the drag are proportional to the square of the speed*. It is one of the fundamental laws of the whole subject.

22. Effects of Size

Both lift and drag also depend on the size of a body; large bodies have more drag than small ones of the same shape; large wings have more lift. Probably everyone knows this too, and it might even be said to be rather obvious, but there is a little more in it than that. From this point of view "size" used to be taken as meaning *frontal area*, i.e. what you see of a body when viewing it from the front—in other words, its cross-sectional area when viewed from this position. For an airship it would mean the area of the largest frame, for a strut the maximum breadth times the length. The greater the frontal area, the greater would be the drag—in direct proportion.

This, however, is another aspect of the subject in which modern development is leading to a change in ideas. When bodies were badly shaped, it was true enough that the frontal area was the best way of thinking of the size of a body moving through the air, but now that so much has been accomplished in the direction of cleaning up and streamlining aeroplane design, now that skin friction has become of so much relative importance compared with form drag, it is more correct to say that resistance is proportional to *surface area* or, as the naval engineer would speak of it, to the *wetted surface*, the surface which is washed by the air passing over it.

Provided bodies are of similar shapes it really makes no difference whether we compare frontal areas or surface areas; for instance, a flat plate two inches square will have four times the frontal area of a flat plate one inch square, and it will also have four times the surface area, and therefore, by both laws, four times the resistance (at the same speed). If, however, either flat plate is faired to form a streamline body, the form drag will, of course, be very much reduced because of the better shape, but we must not forget that there will be an actual *increase in the skin friction* owing to the larger wetted surface and the greater velocity of air flow over it. Think over this, because it is important, and it is apt to be forgotten in view of the decrease in *total* drag. What it means, in practice, is that it may not be worth while polishing a flat plate or a "dirty" aeroplane, but it is very much worth while polishing a perfect streamline shape or a "clean" modern aeroplane, in which skin friction has become the major type of drag. In the case of lift it is usual to consider the *plan area* of the wing.

EFFECTS OF AIR DENSITY

Notice that the area of a full-scale machine is 25 times the area of a one-fifth scale model (Fig. 34) and 100 times that of

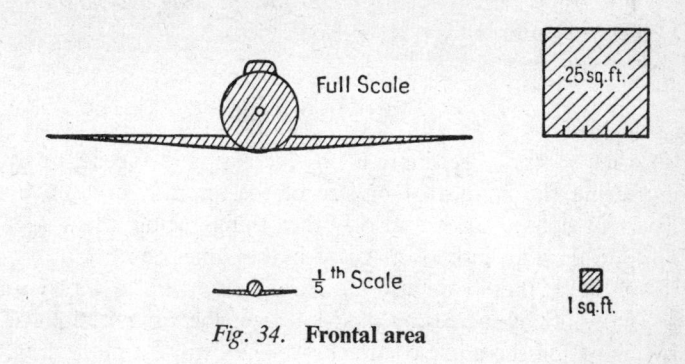

Fig. 34. **Frontal area**

a one-tenth scale model. This applies whether we consider frontal area or wetted surface, or plan area.

23. Effects of Air Density

Lastly, *the lift and drag depend upon the density*, or "thickness" of the air. The denser the air, the greater the forces it produces; this, too, one would expect.

Now, as we noticed when considering the atmosphere, the air density decreases very rapidly as we climb. Even at 20,000 ft (by no means a great altitude for modern aeroplanes) the air density is only about one-half what it is near the ground, and for this reason the drag—other things being equal—should only be half the drag at ground level, so obviously (that dangerous word again!) it will pay us to fly high and thus reduce resistance. But will it? What about the lift? And what about "other things being equal"? That, of course, is where the catch comes in; "other things" at 20,000 ft are far from being equal to what they were near the ground, and it becomes a very debatable question, and a fascinating problem, whether to fly high or to fly low. We shall say more about it later. In the meantime let us remember that *lift and drag depend on the air density—other things being equal*.

24. Lift/Drag Ratio

So when we try to get more lift by increasing the speed, or by increasing the wing area or size of the aircraft, or even by flying in denser air, we also—other things being equal—get more drag, and, moreover, get it in the same proportion; e.g. if we double the lift we also double the drag. But if we try to get more lift by increasing the camber of the wing section, or by increasing the angle of attack, we shall still get more drag, though not necessarily in the same proportion—and this is rather important. The increase in lift is obviously a good thing—the increase in drag is obviously a bad thing—but what is the net result?—good or bad? Of course, there are times when we want lift even at the cost of increased drag (we shall find later that this is so when we are out for low landing speeds); there are other times when we will sacrifice everything, even lift, for a decrease in drag (that sounds like speed records); but in the average aeroplane we shall get a clearer idea of what we are after if we consider the *ratio of lift to drag*, rather than the two quantities separately.

An example will make this clear; the figures are taken from tests on actual wing sections. A certain shape of section gives maximum lift 30 per cent greater than a rather thin section; but, on the other hand, the best ratio of lift to drag of the thinner section is 30 per cent greater than that of the thick section. This is typical of the kind of results which are obtained when wings are tested, and it accounts for the wide variety of shapes of wing section which are in practical use. What it means is that the thicker section would be more suitable for a particular kind of aeroplane, probably a fairly slow weight-carrier or bomber, while the thinner section would suit a more general-purpose machine, and some other shaped section altogether would be needed for a high-speed machine.

Or again, considering the effect of changing the angle of attack of a wing (keeping the speed constant), whereas the lift increases steadily from 0° to about 15°, at which it reaches a maximum, the drag changes very little over the smaller angles with the result that the ratio of lift to drag is greatest (and may be as much as 24 to 1) at about 4°; it then falls off to, say, about half this value at 15° when the lift is a maximum. Of course, once burbling occurs, the lift drops rapidly, the drag increases rapidly, and the lift/drag ratio tumbles to something like 3 to 1 at, say, 20°.

25. Analysis of Drag

Having considered the main factors on which lift and drag depend, let us concentrate for a moment on the unpleasant force—drag.

Why is it unpleasant? Well, lift is what we are seeking; it is what lifts the weight and thus keeps the aeroplane in the air, it makes flight possible, and is the friend of flight. Drag, on the other hand, is a bitter enemy. This backward force contributes nothing towards lifting the aeroplane, and it opposes the forward motion of the aeroplane which is necessary to provide the air flow which in turn provides the lift. This forward motion is produced by the *thrust* and the *thrust* is provided by the *power* of the engine. This applies whether the engine drives a propeller or merely exhausts itself as a jet, or whether the engine is a rocket. The greater the drag, the greater the thrust and the greater the power needed. But more engine power means more weight, more fuel consumption, and so on. and therefore it is fairly clear that for economical flight we must make every possible effort to reduce the drag. So let us analyse it—split it up if we can into its various parts (Fig. 35).

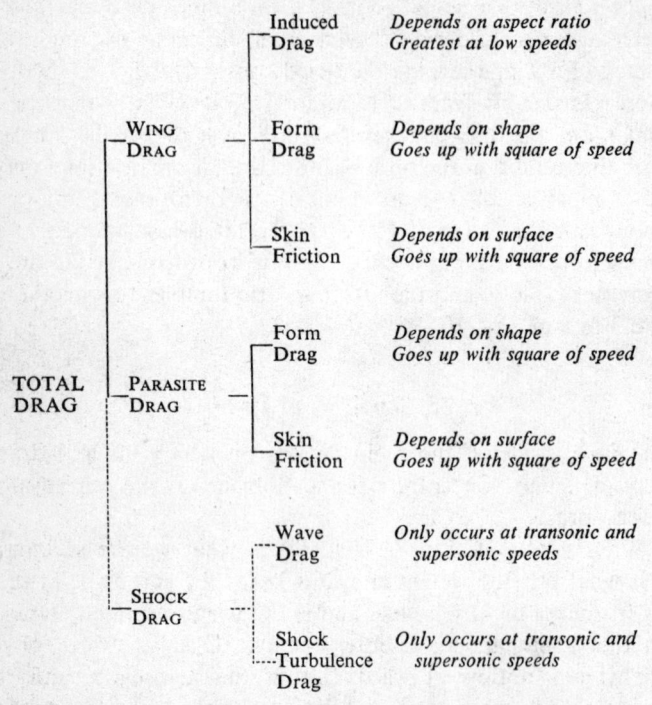

TOTAL DRAG
- WING DRAG
  - Induced Drag — *Depends on aspect ratio / Greatest at low speeds*
  - Form Drag — *Depends on shape / Goes up with square of speed*
  - Skin Friction — *Depends on surface / Goes up with square of speed*
- PARASITE DRAG
  - Form Drag — *Depends on shape / Goes up with square of speed*
  - Skin Friction — *Depends on surface / Goes up with square of speed*
- SHOCK DRAG
  - Wave Drag — *Only occurs at transonic and supersonic speeds*
  - Shock Turbulence Drag — *Only occurs at transonic and supersonic speeds*

Fig. 35. **Analysis of drag**

Shock drag only occurs at high speeds and will be considered in the later sections of the book.

Unfortunately, the drag of a wing is a necessary evil. In the very nature of things, if we are going to deflect the air flow in order to provide lift, we are bound to cause a certain amount of drag. It is true that if the camber is small and the angle of attack is small, the drag will be small—but so will

INDUCED DRAG

the lift. However, it is no good complaining about this, and we become so resigned to this drag from the wings that it has sometimes been called *active drag*. This is rather too flattering a term, but it really implies that it is caused by those parts of the aeroplane which are "active" in producing lift; the term is comparative, it is the lesser of two evils, the greater being its brother of Section 27, and it is really better to call it *wing drag*.

26. Induced Drag

But active drag, or wing drag, the drag of the wings, is in itself made up of various kinds of drag, and the story of the first and most important of these is a fascinating study.

If we tie streamers on to the wing tips of an aeroplane, we shall discover that they whirl round and round as shown in the sketch (Fig. 36). Notice that they rotate in opposite

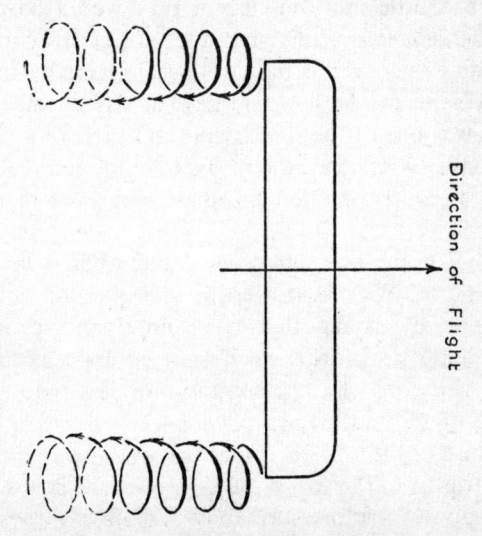

Fig. 36. **Wing-tip vortices**

directions at the two wing tips, the right-hand one going anti-clockwise (when watched from the back) and the left one gonig clockwise. These curious whirls, or *vortices* as they are called, happen with all aeroplanes, but it was a long long time before practical men realized their existence, let alone their significance. What a pity we cannot *see* air; if we could, all pilots from the beginning of flying would have seen, and talked about, these wing-tip vortices; we can easily illustrate them with our piece of cardboard. The author has vivid memories of an incident just after the end of the first war when, on a festive occasion, long streamers were attached to the wing tips of his flying boat. When taxying on the water these streamers rotated violently, and they continued to do so in the air until, after a few minutes, they were nothing but shreds. The author and his colleagues dismissed the whole affair with some such silly remark as "That was funny, wasn't it?" Had they been a little more intelligent they would have realized that a phenomenon of this kind does not occur without good reason, and they would have followed it up by further experiment—and maybe it would have slowly dawned on them that this was one of the most significant facts of aviation and one that was to influence the whole trend of aeroplane design. But that discovery was left to others and, even then, it took some time.

But what is the real significance, and what is the cause of these vortices? We can answer the first question quite simply and shortly by saying that we cannot stir up whirlpools without doing work; this work must be done by the engine, and the whirlpools are nothing more or less than a form of drag tending to hold the aeroplane back.

The cause of the vortices is that the air tends to flow around the wing tip from the region of high pressure below the wing to the region of low pressure above. A fluid always tends to flow from high pressure to low pressure. This flow round

INDUCED DRAG

the wing tips causes all the air over the top surface to flow slightly inwards and that over the bottom surface to flow outwards (Fig. 37). Thus the streams meeting at the trailing

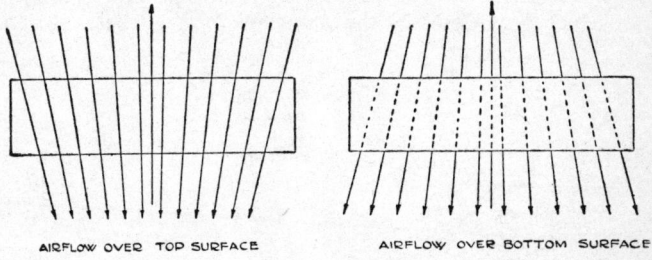

AIRFLOW OVER TOP SURFACE   AIRFLOW OVER BOTTOM SURFACE

Fig. 37. **The cause of trailing vortices**

edge cross each other and form what is really a series of eddies called *trailing vortices*, which roll up to one big vortex at each wing tip (Fig. 38). As a result of the wing-tip vortices

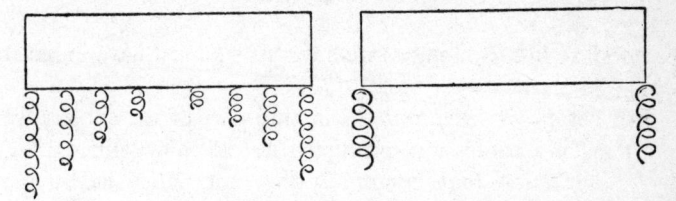

Fig. 38. **Trailing vortices which become wing-tip vortices**

the air behind the wing is deflected downwards, that outside the span being deflected upwards. Thus the net direction of the air which actually passes the aerofoil is in a downward direction, and so the lift—which is at right-angles to the air flow—is slightly *backwards*, and so contributes to what we call the drag (Fig. 39). This is another, and perhaps more scientific, way of thinking of the drag caused by the wing-tip vortices. The drag thus formed is called *induced drag* (another term which the highbrows claim for themselves) because it is a result of the downward velocity "induced" by the wing-tip vortices. In a sense, induced drag is part of the lift, and thus it can never be eliminated, however cleverly we design our wings. This, nuisance as it may be, is really the part of the drag which best deserves the name of "active" because it is

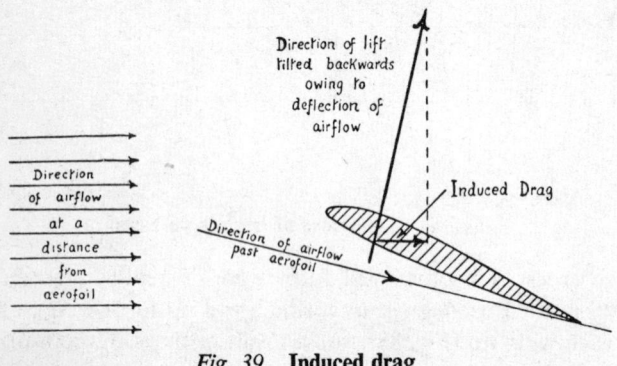

Fig. 39. **Induced drag**

essential to lift. So long as we have lift we must have induced drag.

But before we leave this fascinating part of the subject we must make a confession, prompted not so much, I'm afraid, by a conviction that honesty is the best policy as by the knowledge that we will be found out sooner or later! *Induced drag does not increase with the square of the speed;* on the contrary, *it is greatest when the aeroplane is flying as slowly as it can*, i.e. just before the stalling angle is reached and we are getting the maximum lift for the minimum speed.

27. Parasite Drag

The ideal aeroplane would be *all wing;* it has, in fact, been termed a "flying wing." Even modern aeroplanes often fall a long way short of this ideal; at best they have fuselages, tails,

PARASITE DRAG

and various projections and protuberances, while at worst they are more like Christmas trees than flying wings. These extra parts, engines, radiators, dynamos, guns, bombs, aerials, wheels, petrol tanks, or whatever they may be, all produce drag, but, except in a few instances of clever design, *do not contribute towards the lift*. Their drag, therefore, is considered to be of a very vicious type, and is given the appropriate name of *parasite drag*. The ideal aeroplane would still have a certain amount of active drag, but it would have no parasite drag. We should get better performance; speed, climb, weight-lifting, all would be improved, and at the same time fuel consumption would be reduced. Obviously, therefore, it is well worth the while of those responsible for producing aeroplanes to study this problem of parasite drag, and to see how it can be reduced to a minimum, if not banished altogether.

There are two distinct methods of reducing parasite drag. One is to eliminate altogether those parts of the aeroplane which cause it; the other is so to shape them and smooth their surfaces that their drag is as small as possible. The first is the most effective, but it has its limitations, and progress has been made by trying a bit of each method. The problem of eliminating struts, wires, and projections is really a structural one, and it has largely been solved in modern aircraft (Plates 54 and 55). It is a question of getting strength by internal rather than external bracing, and by having "clean lines" generally.

At one time it was considered that the extra weight required for making an undercarriage *retractable* during flight would be such as to outweigh the advantages which would be gained by the reduction of parasite drag. We do not think like this today; the undercarriage is one of those parts which is useless during flight—worse than useless, it is a parasite spoiling the performance of the aeroplane. Even if it does mean some increase in weight, even if pilots do forget (in spite of various alarm signals) to lower them for landing, the fact is that nearly all modern undercarriages are of the retractable type. The tail wheel has gone the same way or has been eliminated altogether in the tricycle or nose-wheel undercarriage, while radiators were first retracted and then disappeared; the "flying wing" may still be a long way off, but it is a great deal nearer than it was twenty or thirty years ago.

The problem of reducing the drag of those parts which we cannot eliminate forms a fascinating study, so let us now turn our attention to that side of the question.

28. Form Drag

In this age, when even motor cars, railway trains, and ships are streamlined, there is no need to explain what streamlining means; but, perhaps for the very reason that we have become so accustomed to the idea, it is rather hard for us to realize that efficient streamlining took a long time to come, and that even nowadays very few people fully appreciate how effective it is.

The sketches give some idea of the nature of the air flow past bodies of various shapes, and at the same time an indication is given of their comparative resistances (Fig. 40). It

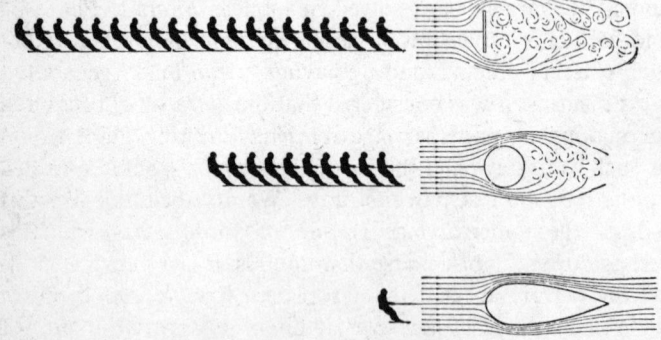

Fig. 40. **The effect of streamlining**

FORM DRAG

will be noticed that the more turbulent the air flow the greater is the resistance, and streamlining really means so shaping a body that air (or water) will flow past it in streamlines, i.e. without eddying, and thus the resistance is reduced to a minimum. Streamlining is another instance in which an attempt to avoid figures altogether would leave us in the dark. How many people realize that by carefully streamlining a flat plate, such as, for instance, a coin held at right-angles to the wind (Fig. 41), we can reduce its resistance not by 20 per

Fig. 41. **Streamlining a coin**

cent or 30 per cent or even 50 per cent, but to less than one-twentieth of its original resistance, a reduction of 95 per cent?

That part of the drag which is due to the shape or "form" of a body, and which can be reduced by streamlining, is called *form drag*.

The sketches show how, in course of time, aeroplanes (Fig. 42*a*), railway locomotives (Fig. 42*b*), motor cars (Fig. 42*c*), and even motor-car lamps (Fig. 42*d*) (until they became incorporated in the body of the car itself, which is better still) were streamlined so as to reduce their head resistance, or form drag. Advantage is often taken of the fact, clearly shown by Fig. 40, that most of the benefit is due to the shaping or "fairing" of the trailing edge of the body and that it makes comparatively little difference whether the nose portion is flat, round or streamlined. This brings back memories of our old friends the venturi tube and the wing section.

Have a look at Plates 54 and 55 again. They give a very good idea of the progress in general cleaning-up that has been made in recent years.

Fig. 42

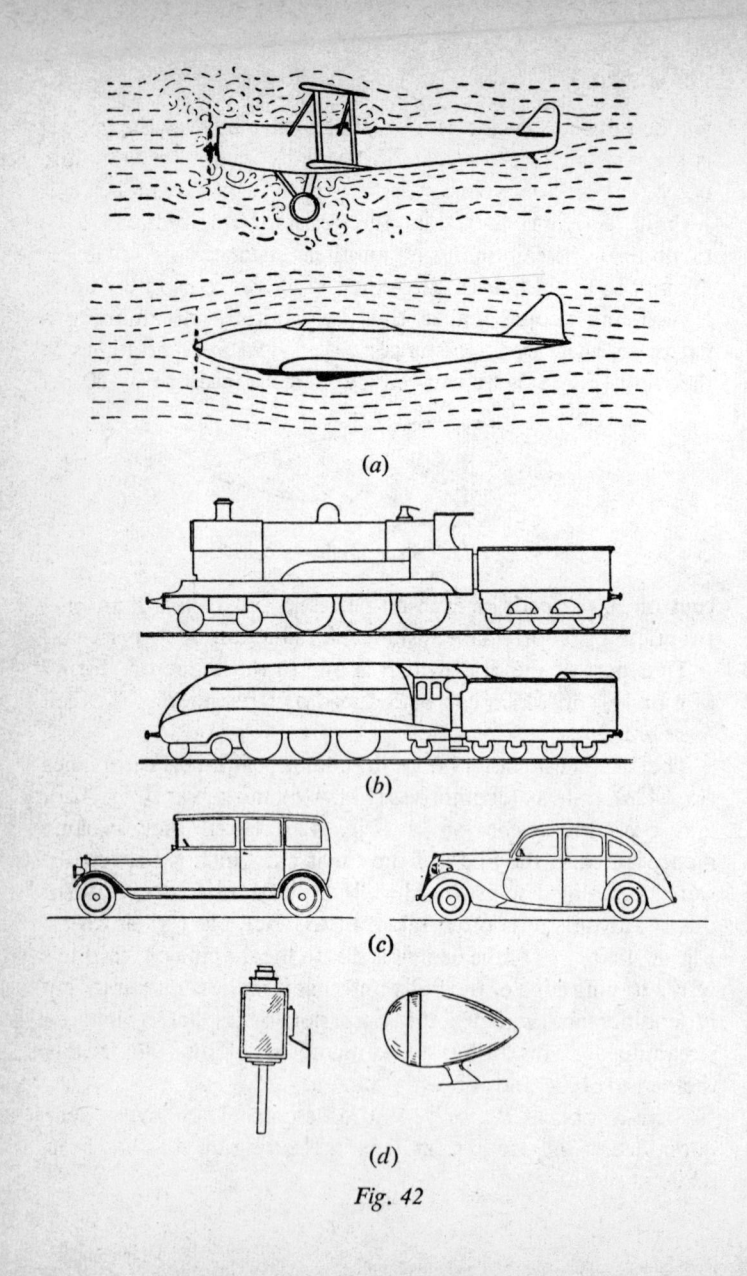

SKIN FRICTION 65

The ratio of *length* (*a*) to *breadth* (*b*), as shown in Fig. 43, is called the *fineness ratio* of a streamlined body. For best results it should be about 4 to 1, but it really depends on the air speed; the higher the speed, the greater should be the fineness ratio, but experiments show that there is not much variation in the drag for quite a large range of fineness ratios.

An aeroplane is made up of various distinct parts such as fuselage, wings, undercarriage and so on. If one could imagine

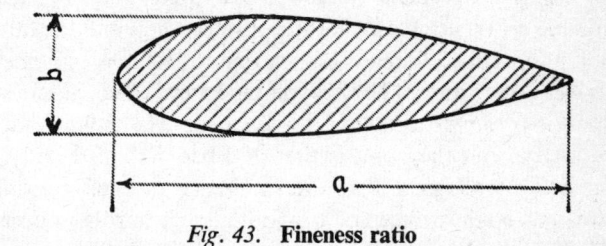

Fig. 43. **Fineness ratio**

each of these parts so shaped in itself as to give the least possible resistance it does not follow that when they are joined together the combination will give the minimum resistance. Resistance caused by the effect of one part on another is called *interference drag*, and much care has to be taken in modern design to reduce this portion of the drag by careful fairing of one shape into another.

29. Skin Friction

Not only the shape of a body, but the nature of its surface also, affects the drag. It can easily be understood that a rough surface will cause more friction with the air flowing over it than will a smooth surface. This surface friction is called *skin friction*. Figures are not so convincing in this case, partly because no parts of an aeroplane are likely to be very rough, and therefore we can only compare a surface like that of ordinary doped fabric with a highly polished metal surface. The former is certainly rough in comparison with the latter; but the difference is not great, and the effect on the total resistance of a highly polished wing surface in place of a fabric surface is not very noticeable—or was not until recently.

For there are two modern tendencies which are making the study of skin friction become of increasingly greater importance. One is *speed*. Whereas at 100 m.p.h. there may be only a negligible difference between the polished metal and the fabric, at 400 m.p.h. the difference is such that it becomes of immense practical importance. The second tendency which affects skin friction is the *improvement in streamlining*. That sounds rather paradoxical, but the point is that in a badly shaped body the form drag is so great that the difference in total resistance between a rough surface and a smooth one is hardly noticeable —it is swamped by the large resistance due to eddies. On the other hand, when a body is so perfectly streamlined that its form drag almost disappears, then the skin friction becomes not only noticeable but important. Clearly, then, high-speed aeroplanes need to have both streamlined shapes and highly polished surfaces. There appears, however, to be a limit to the degree of polish which makes any difference—perhaps that is just as well for those who will be expected to maintain the polish. A surface is said to be *aerodynamically smooth* when further polishing will not have any appreciable effect on its skin friction.

Modern theory seems to suggest yet another reason for the importance of reducing skin friction. Apparently two and two do not make four—in other words, the total parasite drag is not the sum of the skin friction and the form drag, but it is more nearly the greater of the two. Thus by reducing one part of the drag *only*, we do not notice much effect on the *total*. The only way is to reduce *both*.

THE BOUNDARY LAYER

We know another example of this sort of thing in the case of noise. Two equal noises occurring at the same time do not make double the noise; actually they make very little more than one noise. The two chief sources of the noise of an aeroplane are the engine exhaust and the propeller. The former can be silenced, at any rate in piston engines; but it is hardly worth while, because it makes little difference to the total noise.

A part of the drag which might seem to be a necessary evil is what is called the *cooling drag*, i.e. the resistance caused by the air flowing over radiators (in liquid-cooled engines), over cylinders and cowling (in air-cooled piston-driven engines), and through and over turbine engines. This cooling drag is made up of both form drag and skin friction. Much ingenuity has been spent in trying to reduce it; the wings themselves have been used as radiators, and for air-cooled engines special cowlings and ducts have been devised. Results have been good; so good that it has been possible to reduce this portion of the drag to nothing, or even to less than nothing, the heat of the engine being used to help the aeroplane forward. There is nothing miraculous about this; it is simply a little bit of jet propulsion in piston-driven engines and in turbines driving propellers, while in pure jet engines it is, in effect, the *thrust* instead of being drag at all.

30. The Boundary Layer

The study of skin friction has led to an interesting investigation. We have talked about air "flowing over a surface," but probably air never flows over a surface. However smooth the surface may be, the particles of air which are actually in contact with the surface remain stationary relative to the surface and do not move over it. The next layer of air slides over the stationary layer at a small velocity (Fig. 44), the next layer slides over that one at a slightly higher velocity, and so on, until eventually the air is moving at what we would call the "velocity of the air." This region (in which the velocity changes from zero at the surface of the body to the full velocity at the outside) is called the *boundary layer*. Its thickness may be only of the order of one-hundredth of an inch or so; yet, when the rest

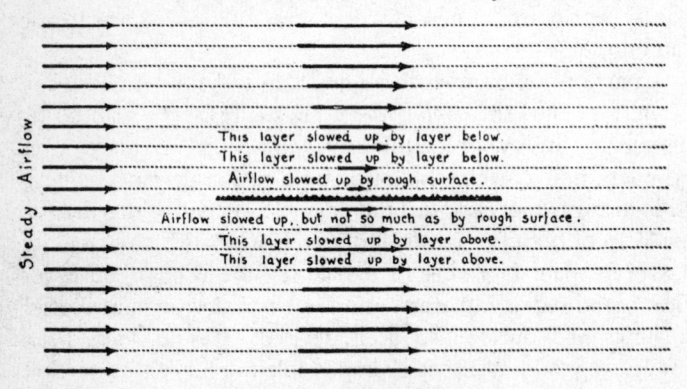

Fig. 44. **Skin friction**

of the air is flowing smoothly, the boundary layer must be in a state of turmoil—called *turbulence*—and thus cause a lot of drag, which is nothing more or less than the skin friction we have been talking about. It is also the break-away of the boundary layer from the surface which leads to stalling. If we can learn to control this boundary layer, to keep it smooth, to keep it close to the surface, and so on, we may succeed in reducing drag considerably. This can be done by having small holes in the surfaces of wings and other parts, and suction inside the wings (Fig. 45), or alternatively by blowing air out and so smoothing the air flow in the boundary layer—the efflux from jet engines has even been used for this purpose.

It is in the boundary layer that the property of *viscosity* of the air is important. It is rather difficult to explain what this

THE BOUNDARY LAYER

term means except by saying that *treacle is very viscous*. It is the tendency of one layer of the fluid to "stick" to the next layer and to prevent relative movement between the two. One can feel this in treacle, one can imagine it in water; but one

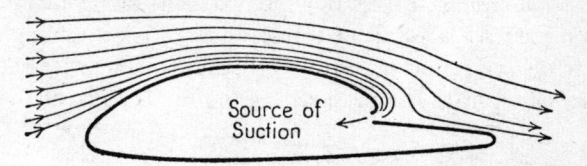

Fig. 45. **Control of boundary layer by suction**

would hardly think of air as being "sticky"—yet sticky it is, though of course to a much less degree than water, let alone treacle. It is this property of viscosity that causes skin friction, and in fact it is ultimately the cause of all turbulence, all eddies and all drag. Yet it is only really effective in this small boundary

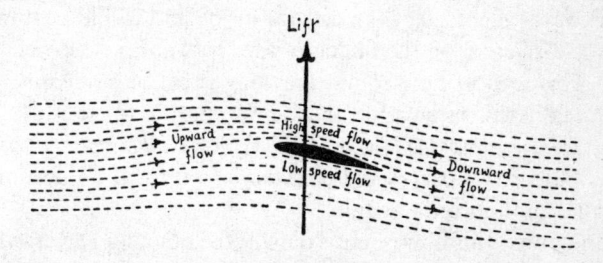

Fig. 46. **Flow of air past an aerofoil**

layer, outside which the air behaves almost as though it were not viscous.

Now that we understand something about the boundary layer and viscosity, we can think of lift from a different point of view. If, by means of smoke or other device, we watch the flow of air over a wing inclined at a small angle, we now know that it will look rather like the flow shown in Fig. 46.

Notice that the streamlines are closer together above the aerofoil than below it, which means that the air must be *flowing faster above the aerofoil and slower below*. Notice also that there is an *upwash in front of the leading edge* and a *downwash behind the trailing edge*. If we could float along in the air stream past the aerofoil, it would almost look as though air was travelling round the aerofoil, because whereas we should be travelling at the same speed as the main body of the air

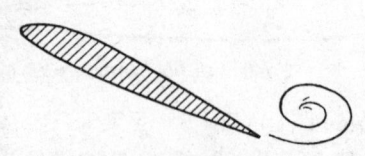

Fig. 47. **Starting vortex**

stream, the air in front of the aerofoil would be moving upwards relative to us; the air over the top would be flowing backwards, i.e. faster than us; the air behind would be flowing downwards and on the under surface forward, i.e. slower than us. This idea of flow round the wing is called *circulation* and it is really this circulation that is responsible for the lift. *But it must not be imagined from what has been said that any particles of air actually travel round the aerofoil*—it is all a question of *relative* motion once again.

Of course, this flow of air is outside the boundary layer, which under such conditions of steady flow is of very small thickness, expecially over the front portions of the aerofoil. Yet, in a way which we cannot properly explain here, it is this thin boundary layer which is responsible for setting up the circulation and so causing the lift. It is interesting to note that, when an aerofoil *starts* to move through the air, the boundary layer causes an *opposite* circulation in the form of an eddy shed from the trailing edge (Fig. 47). Such an idea amuses some people, who think it is a fanciful theory—but it is not, it is a

THE BOUNDARY LAYER

fact, and one which you can very easily see for yourself by moving an inclined surface through water, or even the piece of cardboard through smoke.

When the angle of attack of the wing is increased, the boundary layer becomes thicker and of increasing turbulence, and this turbulence gradually spreads towards the leading edge. Eventually the main flow breaks away altogether from the top surface, large eddies are set up, and stalling, or burbling,

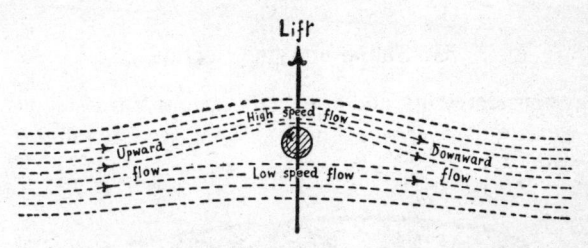

Fig. 48. **Flow of air past a rotating cylinder**

results. Much of the circulation is lost, and so the lift falls off.

Our ideas of circulation become more convincing if we think of a rotating cylinder moving through the air (Fig. 48). The boundary layer will tend to rotate with the cylinder, thus causing an increased speed above and a decreased speed below it (assuming that it rotates in the direction shown in the diagram while it moves from right to left). Also there will be an upwash in front and a downwash behind. In short, we have the same state of affairs as on the aerofoil, and for the same reasons a decrease of pressure above and an increase underneath, and thus a net lift. At first, it sounds a strange idea that a round cylinder can *lift*. It may sound strange, but once again it is no idle theory but a simple everyday fact. If I tell you that it was the principle of Flettner's Rotor Ship, with its large rotating funnels, you may not be much the wiser or the more convinced. But it is much more than that. Do

you play golf, football, tennis, cricket, table tennis or any ball
game? If so, you will know what is meant by putting "top"
or "bottom" spin on a ball, you know how balls are made to
swerve accidentally or intentionally as they travel through the
air. It is all caused by this mysterious lift (notice once again
that lift need not be upwards, but may be sideways or even
downwards); it is all a question of boundary layer and
circulation.

31. Shape of Wing Section

Having considered lift, and drag in its various forms, let us now
see if we can discover what shape of wing section will give the
best results.

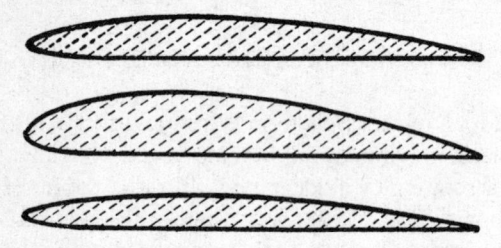

Fig. 49. **Wing shapes with different cambers on upper surface**

Assuming that the wing is to be a double cambered surface,
we still have to decide how much the camber shall be. Fig. 49
shows three typical sections with different top-surface cambers
and so different thicknesses. Generally speaking, a large
camber on the top surface will produce good lift but large drag,
not only induced drag, but form drag; for wings too have
form drag and skin friction in addition to their induced drag.
Different cambers on the under surface do not make so much
difference to the lift and drag properties of the aerofoil, but

the modern tendency has been to change from the very much concave cambers of the early aircraft to much flatter cambers, and even to convex cambers as shown in the diagrams. One would think that cambering the under surface in this way would tend to spoil the downwash and thus affect the lift, and this is to a certain extent true; but, on the other hand, the convex under surface has two advantages which probably outweigh any small loss of lift. In the first place, the depth of the wing is increased, and the deeper the wing the lighter can be its construction; and this reduction in weight is more valuable than the loss in lift. Secondly, the convex under

Fig. 50. **Laminar-flow aerofoil section**

surface has an appreciable effect on the movement of the centre of pressure, tending to make its movement stable, or at any rate less unstable.

Fig. 50 shows another tendency in what is called a *laminar-flow wing section;* notice how thin this section is and how much farther back is the point of greatest thickness than in the more conventional section.

32. Variable Camber

Some advantages result from large camber, others from small camber, and the reader may wonder whether it is not possible to *alter* the camber of a given wing section so as to meet the varying requirements of flight. To do so is certainly a practicable proposition, but it raises a problem which we shall always be coming up against in this subject—whether it is worth while; that is to say, whether we shall gain enough to make up for (perhaps I should say, to *more* than make up for) what we shall lose by the increase in weight of the mechanism involved and the increase in complication. All such devices mean something more to go wrong, some extra lever for an already harassed pilot to worry about.

Many ideas have been suggested, and many ingenious devices patented, in attempts to provide the wing with a "smoothly variable" surface, or even with a variable area. Few of these, however, have got beyond the stage of being ideas, and the only devices that have proved really successful in practice may be summed up under the headings of *slots*, *slats* and *flaps*. These are perhaps more crude than a smoothly variable wing would be, but they have won the day because they combine effectiveness with simplicity—a combination of qualities that is all too rare in modern aircraft but all the more welcome when it can be found.

33. Slots, Slats and Flaps

Flaps at the trailing edge (Plate 42) date back to the First World War, or even before that, but then they were only used on special types of aircraft for special purposes, as for instance on aircraft used in the early experiments in landing on decks of ships. Now, however, flaps are considered to be almost a necessity and, in one form or another, are incorporated in the design of nearly all modern aircraft.

The effect of the trailing edge flap is to increase the camber by lowering the rear portion of the wing, which is made in the form of a hinged flap—similar to an *aileron* (Section 54), except that it probably extends along most of the span of the wing.

The kink thus caused in the top surface may be eliminated by using a *split flap*. In this device the flap portion is split inof two halves, the top half remaining fixed and the bottom half dropping like a lower jaw of a mouth. Many other kinds to special flap have been invented, including some fitted at the

SLOTS, SLATS AND FLAPS

leading edge of the wings, and merits are claimed (by their inventors) for all of them. For some it is claimed that they give the greatest increase in lift, for others that they give the greatest increase in the ratio of lift to drag, and for others that

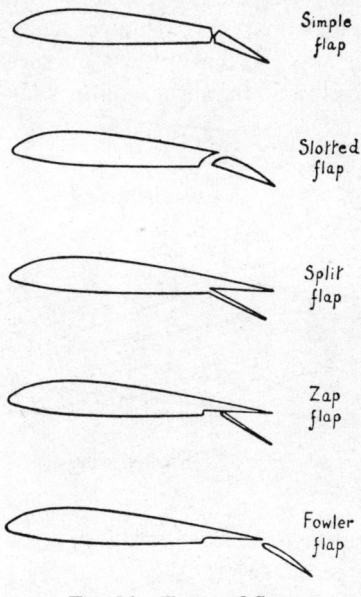

Fig. 51. **Types of flap**

they give the greatest increase in drag. Since all these qualities —even the increase in drag—may be needed for varying circumstances, there may be something in all the claims put forward.

We shall have more to say about various types of flap in a later paragraph (Section 65). In the meantime, have a look at Fig. 51, which illustrates some of the main types.

Slots (Plate 42) have not had quite such a long history as that of flaps, and in view of their early promise, have proved somewhat disappointing. The original object of the slot was to delay the stall of a wing and so obtain greater lift from it.

It has already been explained that the cause of the stall is the airflow breaking away from the top surface and forming eddies. In a slotted wing this is prevented, or rather postponed, by allowing the air to pass through a gradually narrowing gap near the leading edge, so that it picks up speed (a venturi in fact) and is kept close to the surface of the wing (Fig. 52). The

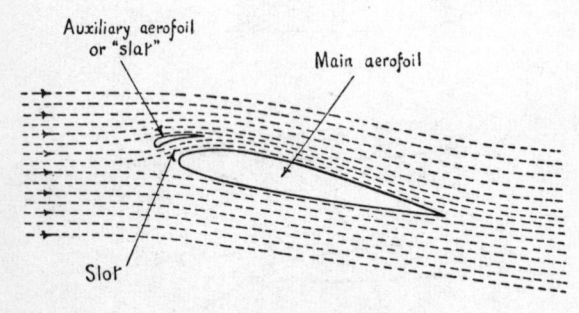

Fig. 52. **Slotted wing**

gap is really the *slot*—the small auxiliary aerofoil which forms the top surface of the gap is called a *slat*.

The effectiveness of slots varies with the type of wing section to which they are fitted; in some instances the increase in maximum lift reached may be as much as 100 per cent, while the stalling angle is increased to 25° or 30°. From many points of view the increased angle is a disadvantage, as will be explained when we are considering landing, and perhaps this has been one of the main causes of disappointment.

Of course slots, like flaps, should be put out of the way when they are not required; otherwise they would tend to cause excessive drag. Fortunately this can be done automatically; at small angles of attack, i.e. at high speed, the air pressure on the

ASPECT RATIO

slat causes it to close while at high angles of attack, i.e. at low speed, the air pressure causes the slat to move forwards and so open the slot. Sometimes, however, slots are controlled by a lever in the cockpit, sometimes they are combined with flaps, and sometimes they remain open all the time.

34. Aspect Ratio

In addition to the cross-sectional shape of a wing, we must consider its *plan* shape, especially the ratio of its *span (or length)* to its *chord (or breadth)*. This is called the *aspect ratio* of the wing. Fig. 53 shows how it is possible to have wings of

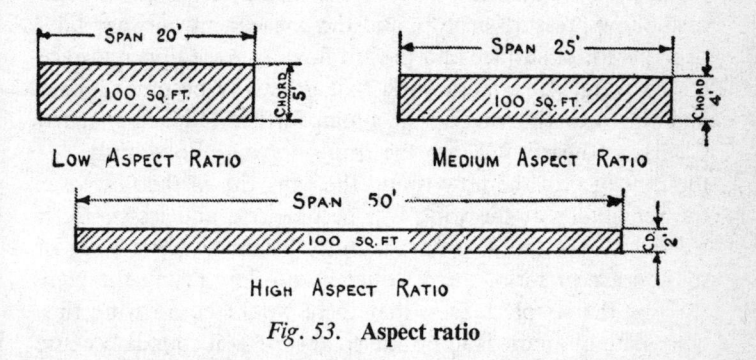

Fig. 53. **Aspect ratio**

the same area but very different aspect ratios. We have said that induced drag cannot be altogether eliminated—because it is an inevitable result of lift. But it can be reduced, even without reducing the lift, and that is where aspect ratio comes in. Experiments indicate that there is a small but quite definite increase in efficiency as we increase the aspect ratio, keeping the area the same. That is why you will notice the very high aspect ratios used on the wings of gliders, sailplanes, and aeroplanes designed for long-distance flying; all cases where efficiency of the wing is of primary importance. But of this we shall have more to say later.

At first it was rather difficult to explain why aspect ratio should be so important, because the elementary theory had led us to believe that the lift of a wing depended on its area, and yet an aeroplane with a high aspect ratio wing was found to be more efficient than an aeroplane with a wing of the same area but lower aspect ratio. The answer to this puzzle is *induced drag*, the wing with the higher aspect ratio having less induced drag.

Why does aspect ratio affect the induced drag? To answer that let us go back to the fundamental cause of induced drag, the flow round the wing tip from the high pressure underneath to the low pressure on top, and the consequent outward flow over the lower surface and inward flow over the upper surface of the wing. Imagine a wing that gradually becomes longer and narrower, the wing tips becoming farther and farther apart. Clearly—I nearly fell into the trap of writing "obviously"!— the influence of the flow round the wing tip on the flow over the remainder of the wing will become less and less until, if we reduce the thing to an absurdity by imagining a wing of infinite aspect ratio, there would be no flow round the wing tips for the simple reason that there would be no wing tips. This state of affairs is not quite so absurd as it sounds because we can, in a wind tunnel, fake conditions of infinite aspect ratio. In a closed tunnel we can do this by making the span of the aerofoil such that it just fits into the tunnel and the tunnel walls effectively prevent any flow round the wing tips; in an open tunnel we can do it even more convincingly by testing a wing of which the span is greater than the width of the jet of air, so the wing tips are outside the jet altogether. Fakes of this kind are, in fact, extremely valuable because they enable us to confirm the theory by taking it to its limits; something that we cannot do in actual flight. As it happens, theory and

ASPECT RATIO

experiment give extraordinarily similar results in this part of the subject, and prove convincingly that the greater the aspect ratio the less is the induced drag.

As so often happens in the study of flight, we find a fly in the ointment—a high aspect ratio has its disadvantages. These are chiefly structural—adding to the weight and thus eventually cancelling out the effect of increased lift. Another bad point is that a high aspect ratio makes a machine more difficult to manoeuvre, whether in the air or on the ground, and it takes up more space in a hangar.

Thus we must compromise on the question of aspect ratio, just as we had to in deciding the amount of camber. Values used in practice vary from 5 or 6 to 1 for fighters, which must be manoeuvrable, to as much as 20 to 1 for sailplanes, but there are certain rather freak examples right outside these limits—in both directions.

We shall mention later the very low aspect ratios of wings used in flight at supersonic speeds, but in the meantime have a look at the Fairey Delta 2 or the Concorde (Plates 23, 32 and 64) and compare their aspect ratios with that of a sailplane (Plate 46).

Before leaving the subject of induced drag—for the time being; we can never leave it altogether—we must once again modify an impression that may have been left by an earlier remark to the effect that it was a long time before practical men realized the significance of wing-tip vortices, and so of induced drag. If by practical men we mean the men who fly, and perhaps even the men who design aeroplanes, then the remark is substantially true, but it is only fair to say that there were other men, the greatest of whom were Lanchester in Great Britain and Prandtl in Germany, who studied, wrote about and preached the principles of induced drag—though they didn't call it that—in the very early days of aviation; it can even be claimed that Lanchester did so before any aeroplane ever flew! When one realizes that those principles explain the importance of high aspect ratio, the advantages of the monoplane over the biplane, and the modern ideas about economical flying, it seems rather hard that in their day these men were not considered as practical men, or even listened to by those who considered themselves to be so. But there it is.

35. Biplanes

And so we come to *biplanes*. It is not easy to discover who first thought of the idea of a biplane, i.e. of using two aerofoils, one placed above the other. Some people, of course, have thought of putting even more planes on top of one another (Plate 6). Many of our ideas about flight have, very naturally, come from birds, but the biplane idea seems to be a purely man-made invention, though some naturalists claim that there are biplane insects. At any rate, the first aeroplane to fly was a biplane (Plate 1), so the idea is at least as old as the history of flight.

We noticed in an earlier paragraph that very large wing areas are required for flight, and the advantage of the biplane was that this large wing area could be arranged in a more compact fashion, making the finished aeroplane more convenient to handle both on the ground and in the air. The biplane structure seemed more suited than the monoplane to give us what we most required: strength without weight. So far the biplane seemed to have all the advantages; why, then, has it proved the loser in the long run?

It is as a wing, as an aerofoil, that the monoplane has always been superior. Remembering how the pressure is distributed round a wing section, let us put two such sections together, one above the other, and observe the effect (Fig. 54). We find that the increased pressure on the under surface of the upper wing is not so effective as it was when it was alone—still less is the decreased pressure above the lower wing so effective; thus both upper and lower wings suffer. There is, in fact, an interference between the two wings and this is called *biplane interference*. Another way of thinking of it is to consider the induced drag, which is greater on a biplane— with its four wing tips—than on a monoplane of the same

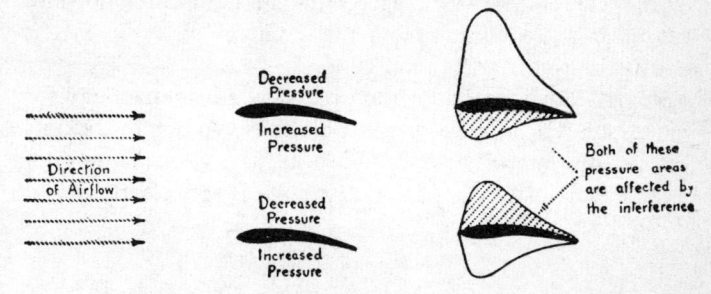

Fig. 54. **Biplane interference**

wing area, and so the overall lift/drag ratio of the monoplane is better than that of the biplane.

The biplane enthusiast, full of confidence owing to the structural superiority of the biplane, persistently endeavoured to minimize this disadvantage.

His first idea was naturally to increase the *gap*, i.e. the distance between the two wings. This expedient had its effect in reducing interference, but very large gaps were needed to make the effect appreciable, and very large gaps meant an increase in structure weight, which, after a limit had been reached, outbalanced the advantage gained.

But our biplane fan was not yet baffled. He next tried to eliminate the interference by *staggering* the planes, in other words separating them horizontally rather than vertically. When the leading edge of the upper plane was in front of the leading edge of the lower plane it was called *forward* or *positive stagger*; when behind it, it was called *backward* or *negative stagger*. Forward stagger definitely served its purpose, and there was a small but appreciable increase in lift when compared with an unstaggered biplane of the same gap. The picture of the B.E.2C (Plate 3) shows clearly the large gap and forward stagger which was so common at the time of the First World War. Backward stagger, although it appeared hopeful, was most disappointing from this point of view; in fact it actually did more harm than good.

Stagger, however, had certain practical advantages, and for this reason was adopted on most biplanes. Access to cockpits was usually improved, and, above all, the view of the pilot became more extensive. This latter point is very clearly shown in Fig. 55.

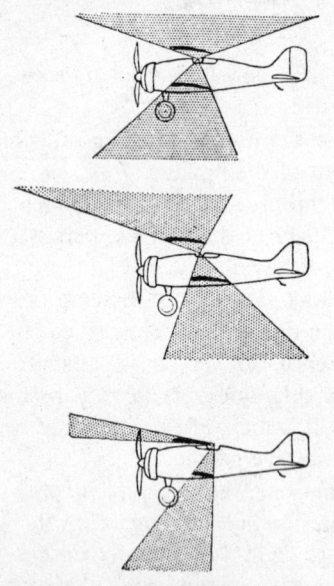

Fig. 55. **Angles of view**
The shaded areas shows the blind spots

The reader who is not used to flying may not realize the seriousness of this question of field of vision. On the other hand, the reader who flies frequently has probably become so accustomed and resigned to seeing only a little less than nothing during flight that he does not realize how many "blind spots" there are in the average aeroplane—whether biplane or monoplane.

Anything that can be done to enlarge the pilot's field of vision is a step in the right direction, and may well have turned the balance in favour of stagger.

The *sesquiplane*—or one and a half plane—was really a compromise between a monoplane and biplane. The reader may have noticed that we are frequently using that word "compromise"; no wonder, because it crops up in every part of aeroplane design. A finished aeroplane *is* a compromise from beginning to end. We want this, we want that; but we cannot have both this and that, so we end up by having a bit of each. The sesquiplane was a bit of a monoplane and a bit of a biplane. The structure was that of a biplane and had its consequent advantages; on the other hand, the lower plane was so small that it caused hardly any interference with the upper plane, which was therefore "almost a monoplane." Plates 5 and 7 illustrate this tendency towards a large upper plane and small lower plane.

But even in a sesquiplane there were struts and wires to connect the two planes, and when, further to tip the balance, experience in structural design and the improvement of structural materials, together with other advances in aeroplane design, made it possible for a monoplane structure to be as efficient as that of a biplane, designers came slowly but surely round to the opinion that the monoplane was the best type. So perhaps the birds, not to mention Lanchester and Prandtl, were right after all.

36. Lift and Drag—A Summary

We have so far considered the forces that act upon bodies due to their movement through the air, and how they experience lift, or drag, or both, according to their shape, speed, and so on. We are now in a position to study something even more interesting—the flight of the aeroplane as a whole—but, before doing so, let us sum up what we already know about lift and drag:

(a) A body that is pushed or pulled through the air causes a disturbance in the air and, in consequence, experiences a force.

(b) The amount of this force depends on the shape of the body,

(c) on its speed through the air (actually, speed squared),

(d) on its size,

(e) on the smoothness of its surfaces,

(f) and on the density of the air through which it passes.

(g) That part of the force which is parallel to the direction of the air flow, that is to say which acts against the motion of the body, is called drag.

(h) That part of the force which is at right-angles to the direction in which the body is travelling is called lift.

(k) A wing is designed to give as much lift as possible with as little drag as possible.

(l) Other parts of the aeroplane, if they cannot be eliminated altogether, are designed to give as little drag as possible—the drag of these parts is called parasite drag.

(m) Drag caused by the shape of a body is called form drag—this is reduced by streamlining.

(n) Drag caused by roughness of surface is called skin friction.

(o) Wings also experience induced drag, as an inevitable consequence of their lift.

STRAIGHT AND LEVEL FLIGHT

(*p*) The wing is pushed or pulled through the air at a small angle called the angle of attack or angle of incidence.

(*q*) This motion produces a downwash which in turn causes the upward reaction or lift.

(*r*) As the angle increases the lift increases up to a certain angle called the stalling angle.

(*s*) Wing sections are curved or cambered, usually on both top and bottom surfaces.

(*t*) The decrease in pressure on the top surface is caused by the speeding up of the flow over that surface—as in a venturi tube.

(*u*) Slots and flaps are the most practical means of producing variable camber.

(*v*) The top surface of a wing contributes more lift than the bottom surface, the front portion more than the rear portion.

(*w*) The centre of pressure is therefore well forward.

(*x*) As the angle changes, the centre of pressure may move in a stable or unstable way—with most aerofoils the tendency is unstable.

(*y*) Wings of high aspect ratio are the most efficient, because they have less induced drag.

(*z*) After a long struggle the monoplane has won the day over the biplane.

Yes, we have exhausted the alphabet and, what is more important, we have already learnt the main principles on which flight depends.

37. Straight and Level Flight

Let us now apply these principles to the flight of the aeroplanes as a whole. This is where it all becomes more interesting; it is where the practical men, namely those who build and those who fly aeroplanes, will begin to see what we have been leading up to.

The flight of an aeroplane may be divided into two parts:

(*a*) Straight and level flight
(*b*) Manoeuvres

When considering *straight and level flight* we shall include only that rather rare condition in which the aeroplane is moving forward at constant air speed and neither losing nor gaining height. Under the second division—*manoeuvres*—we shall include taking off, landing, climbing, gliding, turning, looping, and spinning—all these, and anything else that an aeroplane may do when it is not in level flight.

It will be clear that the first is the smaller category; but it is none the less of great importance, because it is the basis of aeroplane design. We consider first the loads upon an aeroplane in straight and level flight, then we consider all the other kinds of flight chiefly *in comparison with straight and level flight*: the loads are so many times more, or so many times less, than in straight and level flight.

To the pilot, too, straight and level flight has a significance in that it is one of the first exercises that he learns in the air, and thereafter it remains the basis of all other exercises. Strange as it may seem, however, except on long-distance flights, pilots do not often indulge in straight and level flight—it is rather a dull state of affairs unless one's sole object is to get from one place to another, and even then quite a large proportion of many flights are spent in climbing or descending.

38. The Four Forces

Now, an aeroplane is kept up in the air, and travels through the air, by means of the various forces which act upon it. We have already mentioned the *lift* of the wings, and also the *drag*

THE FOUR FORCES

both of the wings and the parasite variety. The other important forces are the *thrust* which produces the forward motion, and the force of gravity, i.e. the *weight* of the aeroplane.

These forces, as for instance the drag, are made up of many separate parts: there is the drag of the wings, of the wheels, of the fuselage, and of all the other parts—they can all be added up and produce one large drag. Similarly the weight is the added weight of all the parts, the lift may come from two

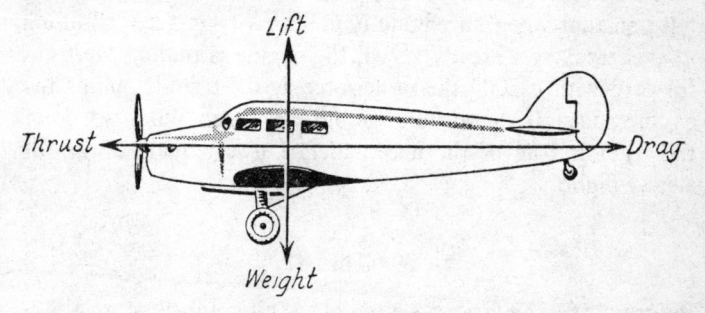

Fig. 56. **The Four Forces**

wings and perhaps also from the tail plane, the thrust from the several blades of a propeller, or possibly from two or more propellers or jets. But we can make a total of each one. Having done so, the forces must be *balanced*, they must be in *equilibrium*, if we wish to maintain a condition of steady flight.

Now, the four main forces act in different directions during straight and level flight. The lift will be vertically upwards, the weight vertically downwards, the thrust horizontally forward, the drag horizontally backwards (Fig. 56).

To maintain a constant height the two vertical forces, lift and weight, must balance, must be *equal*. If you have not learnt mechanics you may be inclined to think that the lift ought to be greater than the weight, ought to beat it, overcome it. If you feel like this, ask someone who understands mechanics to convince you; he will probably have a difficult job, but he ought to succeed in the end—because he is right. The lift of the wings, then, must equal the weight, i.e. the weight of the whole aeroplane.

Now for the two horizontal forces—thrust and drag. Here you are going to give the expert in mechanics an even more difficult job, because he has got to convince you that these, too, are *equal*—the thrust is equal to the drag. He has got to tell you that when an engine is pulling a train forward along a level track at a steady speed, the engine is pulling the train forward with exactly the same force as the train is pulling the engine back. It sounds strange but it is true. What is true of the train is true of the aeroplane. Thrust equals drag—for steady flight.

39. Thrust

Before going any farther we ought to think a little more about the important force, *thrust*. We can only tackle the subject very briefly here, and fortunately the reader who is interested in how the thrust is produced has only to turn to a companion book in this series—aptly named *Thrust for Flight*—in which he will find all he wants to know about it.

All we need say here is that there are several methods of providing this force in aeroplanes, and they all depend on the principle of *pushing air, or something else, backwards with the object of causing a reaction, or thrust, in the forward direction*. If you stand on a slippery surface and throw something away from you, you will begin to slide in the opposite direction; if you continue to throw things, you will continue to slide; the more things you throw, the heavier they are and the faster you throw them, the greater will be the *thrust* that you produce.

Now, there is an obvious practical difficulty about this, the provision of something to throw. If the process of throwing things away is to continue, it is going to need an awful big store in which to pack them before you start, and they are going to weigh a mighty lot. In a ship we get over this by using the surrounding water; the screw, the paddle wheel and even the oar, are each designed to get hold of some of this water, and throw it backwards and thus produce the force that propels the ship forwards. The only substance that is available in large quantities near the aeroplane is the air itself; it is not by any means ideal for the purpose but it is better than trying to carry anything in the aeroplane.

40. Jet Propulsion

Any means of propulsion which depends on throwing the air backwards is really a form of *jet propulsion*, and in that sense all power-driven aeroplanes are jet propelled. But by common

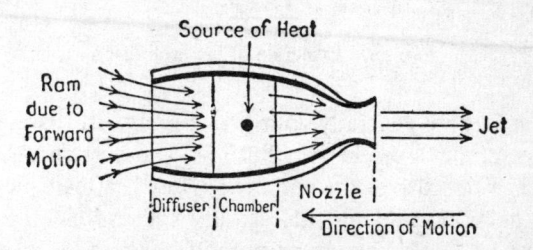

Fig. 57. **Principle of the ram-jet**

usage the term "jet propulsion" has come to be applied only to those forms in which there is no external propeller, the air simply flows into some kind of engine where it is compressed, partly by the natural ram effect and partly by the engine itself, then heated with the aid of some fuel, and having thus acquired extra energy rushes out at the other end faster than it came in.

That is all there is to it, and that is one of its great virtues—simplicity (Plates 16, 25 and 26 and many others show clearly the jet engine nacelles).

The engine itself may be of various types ranging from the pure *ram-jet* (Fig. 57), in which there is really no engine at all but merely a source of heat, to the *Whittle type engine* (Fig. 58),

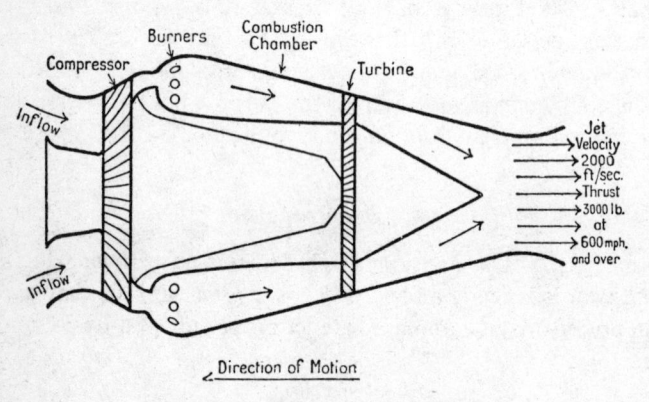

Fig. 58. **Principle of jet propulsion**

or, as it is now generally called, *turbo-jet*, with its collector, compressor or supercharger, burners, combustion chambers, turbine (which drives the compressor) and jet exit pipe. But we are going outside our subject; it is rather difficult to avoid doing so when considering jet propulsion, because engine and aeroplane tend to become one and the same thing (Plate 20 is a good illustration of this tendency).

41. Propeller Propulsion

In *propeller propulsion* (Fig. 59), the "jet" of air which is thrown backwards does not go through the engine itself; it is a cold jet (not having received heat from the engine), much larger in diameter and incidentally having much less velocity than that used in so-called jet propulsion. The rotating propeller blades act very much like the wings of an aeroplane, but we will deal with them in a later section; all we need realize at this stage is that they produce a backward flow of air, called the *slipstream*, of about the same diameter as the propeller itself. The propeller may be driven by any kind of

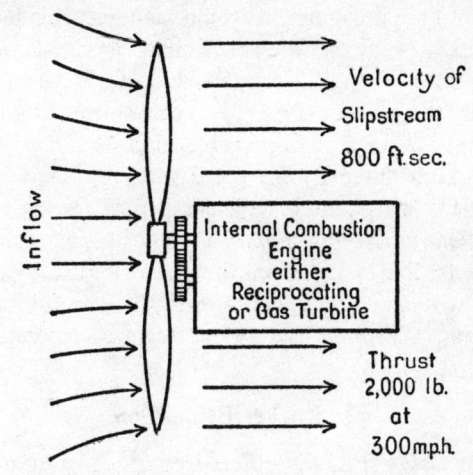

Fig. 59. **Principle of propeller propulsion**

engine, and although in the past it was usually of the reciprocating type, turbine engines are being used more and more for driving propellers (this is called *turbo-prop* propulsion), and it is no longer appropriate to associate propeller propulsion with the reciprocating engine alone.

Propeller propulsion and jet propulsion may be combined, and the turbine engine lends itself particularly to this idea. The exhaust from the turbine forms the jet propulsion, the hot high-speed small-diameter jet; the slipstream from the propeller, the cold low-speed large-diameter jet. We shall see later that there are advantages in both types of jet, and so the combination may be very effective especially since we can allot the proportion of each more or less as we wish when designing the engine (Plate 57).

An interesting aspect of propeller propulsion is that it seems to offer the most hopeful means of the achievement of real man-powered flight—indeed, with the possible but unlikely exception of flapping wings, it would seem to provide the only hope. We say "real man-powered flight" because, of course, aeroplanes have already been persuaded to leave the ground by the exertions of man (Plate 59), but we are still a long way from the time when a man-powered aircraft can be used like a bicycle. That day can only come when we have developed more efficient wings, more efficient propellers, and perhaps more efficient men—but come it will, like so many other achievements that have been declared impossible. In the meantime it remains an exciting dream of the future, one of the few things that man has not yet learnt to accomplish.

42. Rocket Propulsion

Rocket propulsion (Fig. 60) differs from both jet and propeller propulsion in the one important respect that "the things that are thrown backwards" are carried in the rocket itself; in other words, we do not rely on the surrounding air to provide the jet. This fact points immediately to the inherent disadvantage of the rocket system; as we put it earlier "it would need an awful big store." In practice it is just impossible to provide the storage or to carry the weight that would be required for a flight of any reasonable duration, and so rockets may be considered as boosts for other forms of propulsion, to be used only in special circumstances, or for flights whose duration is measured in minutes and seconds rather than in hours (Plates 45, 47 and 58).

ROCKET PROPULSION

Rocket propulsion is simply a development of the firework type of rocket with which everyone is familiar. The fuel or fuels are carried in the "engine," which is nothing more or less than a suitable chamber; when they are ignited, or maybe simply united, they burn furiously and the products of combustion, gases and so on, are ejected by the pressure produced at high speed through a suitable nozzle. Many different kinds

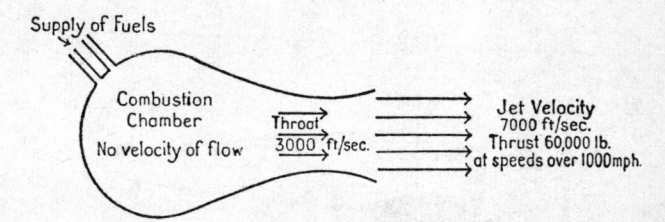

Fig. 60. **Principle of rocket propulsion**

of fuel can be used, solids, liquids and gases; and the system may even lend itself eventually to the use of atomic power.

In emphasizing the inherent disadvantage of rocket propulsion we must not forget the inherent advantage which arises from the same fact that we carry in the aircraft the things that are thrown backwards. This means that we shall get thrust even when there is no surrounding air, and this in turn means that we can fly where there isn't any air, and this—well, you can imagine the rest. No, it isn't quite as easy as all that; no air means no drag, so we won't need any thrust anyway; no air means no lift, so we shall have to use the thrust to balance the weight; but perhaps we are going outside our subject again, because no air means no aeroplanes—we are in the realm of missiles, satellites and spaceships, and that is where rocket propulsion becomes the only practical means of providing thrust—that is where we must hope for another book in this series, because flight outside the atmosphere is an entirely different subject (Plates 45 and 47).

43. Balance of Aeroplane

To return to earth—and to the balance of the aeroplane—the equality of lift and weight as established in Section 38 maintains constant height, and equality of thrust and drag maintains

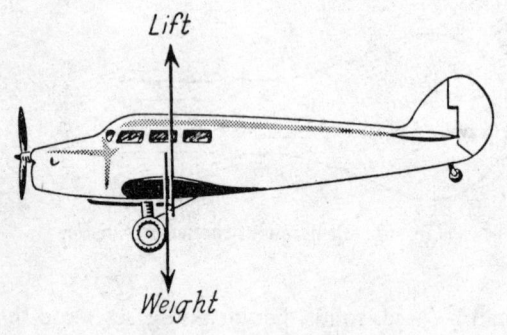

Fig. 61. **Nose heavy**

constant speed. But the *position* of these forces, their *lines of action*, must also be considered. If the lift is far back and the weight well forward, the aeroplane will tend to turn on to its nose—to be *nose-heavy* (Fig. 61). If the thrust is high and the drag low, there will be the same effect. Under the contrary conditions, in both cases, the aeroplane will tend to be *tail-heavy*.

Now, it is part of the job of the designer to see that the aeroplane is neither nose-heavy nor tail-heavy when in straight and level flight, and he must so arrange his forces as to achieve this state of affairs. It is very easy to say "arrange his forces," but what exactly does that mean?

Let us consider the four forces in turn.

The *lift* will come mainly from the wings and so the position of the wings will determine the line of action of the lift force. We must remember, of course, that the centre of pressure is well forward of the centre of the wing and that it moves about with change of angle. Tail plane, fuselage and other parts may each provide a small share of the lift and may influence its line of action, but their effect will be small, and, generally speaking, we can say that the position of the wings determines the line of action of the lift.

The line of action of the *weight* will, of course, be through the centre of gravity. Yes; but whereas it is easy to measure the position of the centre of gravity of a finished aeroplane, it is not easy to estimate it during the early stages of design, especially since we are faced with the fact that if we move, say, the wings to adjust the position of the lift, we shall, by moving the weight of the wings, also alter the position of the centre of gravity. Another problem is to allow for the effect of movable or consumable parts of the weight, such as petrol, oil, bombs, passengers, etc. The aeroplane should be just as well balanced when nearly all the petrol has been consumed as when the tanks are full.

While there may be practical difficulties in arranging for the line of *thrust* to be in the position in which we should like it to be, it is at least the only one of the four forces whose position is obvious to practical men, since its line of action must be along the propeller shaft or centre-line of the jet.

The line of action of the *drag* force will depend upon the positions and relative values of the drags of all the separate parts. In many ways it is the most difficult of all to determine. Unlike the lift, it does not come principally from one particular part, such as the wings; unlike the weight, it does not lend itself to practical measurement; and, unlike thrust, it is not obvious to the eye. Fortunately, however, its possible range of position is not likely to be very great, and any error in calculating its position or changes in its position will not be so serious as a similar error in regard to weight or lift.

In case the reader should not understand the justification for the last sentence, it may be as well to point out that, although —for proper balance—lift must equal weight, and thrust must equal drag, there is no equality needed for the two pairs of forces. The former pair, lift and weight, are greater, many times greater, than the latter pair. Were it not so, the aeroplane would be inefficient. After all, the purpose of the aeroplane is to lift *weight*, as great a weight as possible, with small power— in other words, with small *thrust*. An efficient aeroplane may have great weight but will need only a small thrust; it will have a large weight/thrust ratio, which of course is the same thing as a large lift/drag ratio, because lift equals weight and thrust equals drag. In practice the lift-weight pair may be from five to ten or more times as great as the thrust-drag pair. An efficient aeroplane of 10,000 lb weight may only need a thrust of 1,000 lb to keep it in horizontal flight. Looking at it the other way, a lift of 10,000 lb is obtained for a drag of 1,000 lb (see Plate 43). It will be remembered that in Section 24 we even talked of a lift/drag ratio of 24 to 1; but that was for a wing only, a flying wing, with no parasite drag. It is not possible to reach such efficiency on a complete aeroplane, though some sailplanes with high aspect ratio and clean lines do approach it.

Now, under what conditions will the four forces keep the balance of the aeroplane without causing rotation? One solution is that the lines of action of all four should pass through one point. Then—given the equality of lift and weight, and of thrust and drag—there would be complete equilibrium and no tendency to rotate. This condition was illustrated in Fig. 56.

But quite apart from the practical difficulties which may arise in arranging for this state of affairs, in some cases it may

BALANCE OF AEROPLANE

even be inadvisable to do so. The reason is that in any emergency, such as a sudden stoppage of the engines, it is nearly always preferable for an aeroplane to become nose-heavy rather than tail-heavy. A slight nose-heaviness will tend to put the aeroplane into a natural gliding angle at a time when the pilot has many things to think of and quick decisions to make. Tail-heaviness, on the other hand, would tend to cause

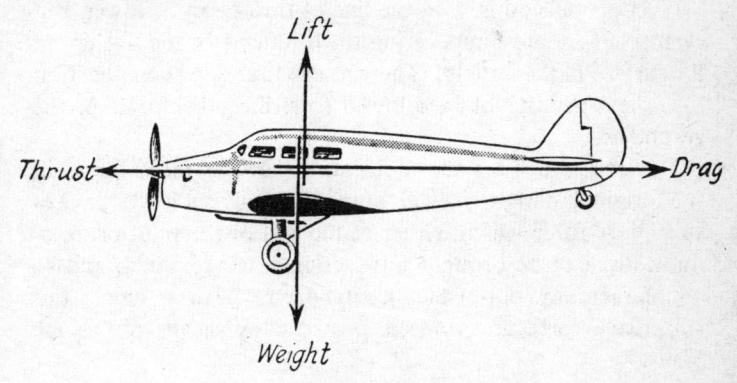

Fig. 62. **Balance of the four forces**

a stall and before he could do anything else the pilot would have to put the aircraft into a gliding attitude.

The best way of making an aeroplane nose-heavy when the engine is stopped, i.e. when there is no longer any thrust, is to arrange for the line of action of the lift to be behind the weight. Very little is necessary, only a matter of an inch or two in a full-size aeroplane. This was shown in Fig. 61.

If we do this, we must take steps to *prevent* nose-heaviness during horizontal flight, i.e. when the thrust is in action. This can be done by arranging for the line of thrust to be low and the line of drag to be high. Fig. 62 shows this arrangement, which is one that is commonly adopted where practical considerations allow it. Since the forces of thrust and drag are smaller than those of lift and weight, they will have to be farther apart to produce the same effect, and it may be inconvenient, or even impossible, to arrange things as we should wish. For instance, if propellers are used we are limited in lowering the line of thrust by the diameter of the propeller, which must clear the ground even when the aeroplane is taxying with its tail well up.

If jet propulsion is used the line of thrust can be lower, but even so, there are limits owing to the effects of the hot jet on the surface of the airfield. There is also the added complication that the jet must not be allowed to strike other parts of the aeroplane.

However much we try to balance the four main forces in the original design, it will always be advisable to have a subsidiary force which, either at the will of the pilot or automatically, can be brought into action when required, and so counteract any out-of-balance tendencies. In practice, this subsidiary correcting force is provided by means of the *tail plane*.

44. The Tail Plane

The Americans call the tail plane a *stabilizer*—and it is a very appropriate name. Its function is to provide a force of such a size and in such a direction as to correct any out-of-balance effects of the four main forces (Fig. 63). The tail plane force need only be small to be effective, because, being situated at the end of the fuselage, it has great leverage. Clearly, an upward force on the tail will have a nose-heavy tendency; a downward force, a tail-heavy tendency.

In the early days of flying, tail planes were usually fixed at some angle so chosen as to give the aircraft correct balance under normal conditions of flight; this meant that aeroplanes were apt to become very nose- or tail-heavy under other

THE TAIL PLANE

conditions, as for instance in climbing or, more particularly, after dropping bombs or when much fuel had been consumed. So acute did this difficulty become that the tail plane was eventually made adjustable (towards the end of the 1914–18 war); that is to say its angle could be altered by the pilot during flight. This was such an improvement that it was rapidly adopted on all aeroplanes. But then, as so often happens, a

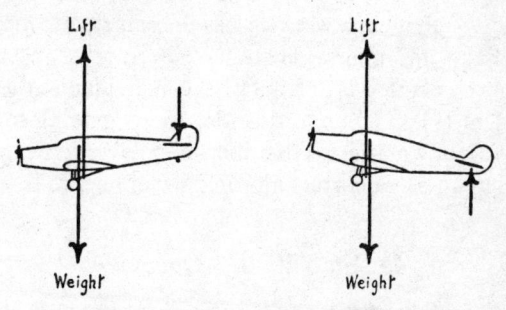

Fig. 63. **Action of the tail plane**

new idea came along. This was the *control* or *trimming tab* fixed at the trailing edge of the elevators, and it will be described in a later section in connection with control. These trimming tabs proved so effective that they caused the adjustable tail plane, in its turn, to become obsolete—or so it seemed—and no one showed many signs of regret at its parting, because a fixed tail plane is very much lighter in construction and can be made of cleaner design and better faired into the fuselage and the remainder of the tail unit; which may make it seem all the more strange that the adjustable tail plane has returned—but so it has, and even to some very modern aeroplanes.

At the rear of the tail plane are hinged the *elevators*, with which the pilot can *upset* the balance at will and thus manoeuvre the machine into any desired position. On some modern types of high-speed aircraft, instead of having a separate elevator, the whole tail plane is designed to move, forming what is called a *slab tail plane* (Plate 21)—another return to an old idea.

There are some types of aeroplane in which there does not appear to be any tail plane in the ordinary sense of the word: these types have sometimes been called tail-less, though actually they have two tails instead of one, the wings being swept back so that each wing tip has the same correcting effect as an ordinary tail plane. There have been other instances in which the tail plane was in front of the main planes, the *canard* arrangement (Plate 63), and this idea too is tending to return, for all that really matters is that there should be some means of introducing an extra correcting force when required.

45. Stability of Aeroplane

Having ensured that the aeroplane will be properly balanced in normal horizontal flight, our next job is to make it stable, i.e. to arrange that, if it is disturbed from its path of flight, *it will automatically return to it again without any action on the part of the pilot*. Fig. 64 illustrates the difference between stable and unstable equilibrium for a body resting on a table, the bottom picture representing a sort of half-way house between the two states. An aeroplane, if disturbed from the equilibrium of steady flight, may exhibit similar characteristics according to its stability or instability.

The reader who is not used to flying may wonder how and why an aeroplane should be disturbed from its straight forward flight unless the pilot deliberately uses his controls, as there are no waves in the air to toss it about or upset it as there are on the sea. No *visible* waves, perhaps; but, as any pilot—or passenger—knows, the air is often in a very disturbed and turbulent condition, and when this is so, it may well rival the

STABILITY OF AEROPLANE

roughest of seas. The chief causes of disturbances in the air are sudden changes of temperature, such as when there is a hot sun and drifting clouds. All kinds of convection currents are then set up, the hot columns of air rising, sometimes with

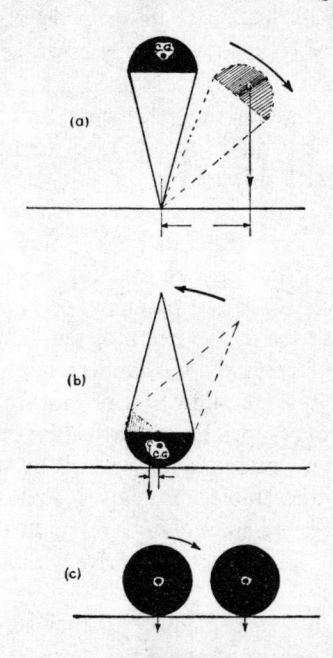

Fig. 64. **Stability and instability**

(a) Unstable equilibrium.
(b) Stable equilibrium.
(c) Neutral equilibrium.

considerable velocity, colder masses of air rushing in to take their place, and somewhere or other there must be downward currents of cold air to keep up the circulation. Similar effects are caused by the wind flowing over uneven ground, the sun shining on mixed spaces of land and water, or on different types of ground such as woods, towns or open countryside—all these absorb the heat differently and currents in the air are set up. Such combinations of circumstances make the air *bumpy*, and when flying through bumpy air the aeroplane is apt to be thrown off its course, to drop the nose, tail, or one wing, or even to be transported bodily upwards or downwards on the currents of air.

In such circumstances the pilot may use his controls to restore the aircraft to its proper path of flight, but although the matter was much disputed in the early days of flight, it is now considered advantageous that the aeroplane should *of its own accord* tend to return to its path. Pilots of today often wonder why it was that anyone ever favoured unstable aeroplanes. It is not very easy to explain, but perhaps a comparison may be made with the introduction of self-changing gears for motor cars. There certainly was, and perhaps there still is, a prejudice against this simplification of motoring, so much so in fact that its progress has been retarded by popular opinion. But why is popular opinion against it? Why was the opinion of pilots against the stable aeroplane? Probably because if one can do a thing, one likes it to be difficult. With self-changing gears even women can learn to drive without—— Well, let us leave it at that!

Although it is now accepted by all that aeroplanes should possess a certain degree of stability, *too much stability is not advisable*. This last remark applies especially to fast fighting aeroplanes, which must be manoeuvred quickly, and whose line of flight must be constantly changed. A very stable aeroplane resists every change in the path of its flight, and it may positively refuse to perform certain useful manoeuvres such as a spin or steep side-slip. On the other hand, an air liner normally engaged in straight flying over long distances should possess a high degree of stability, but even so, not so much as to overrule the pilot's control.

DEGREES OF STABILITY

46. Degrees of Stability

If an aeroplane is disturbed from its path, it is very rarely that it will return to it without some *oscillation* about its original position. There are at least five ways in which it may behave (Fig. 65):

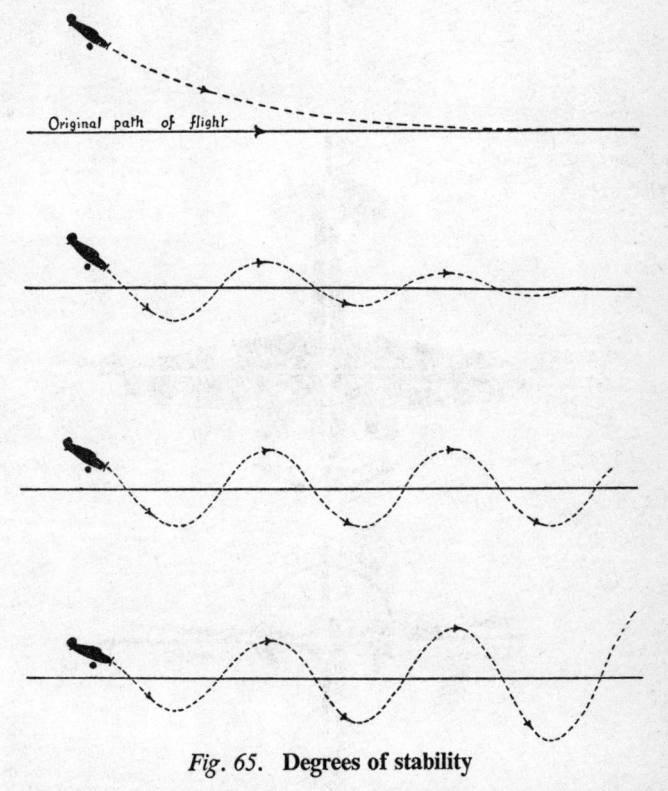

Fig. 65. **Degrees of stability**

(*a*) It may go straight to its path without any oscillation. This is called a dead-beat motion, and does not often occur in practice.

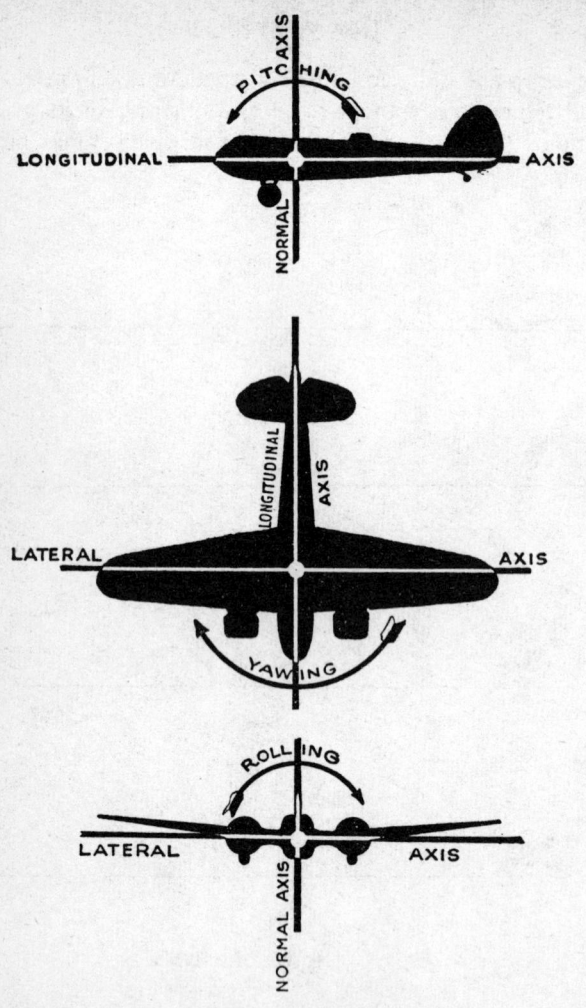

Fig. 66. The three axes

(b) It may oscillate to and fro about its original position, the oscillations being gradually damped out. This gives complete stability of the usual kind.

(c) It may stay in its new position, or it may oscillate to and fro, the oscillations remaining always of the same amplitude. This is a kind of neutral stability—neither one thing nor the other—and is rare in practice.

(d) It may oscillate to and fro, the oscillations becoming steadily greater. Although this is sometimes called stability, it is of a very undesirable type. It may happen in practice and needs watching in design.

(e) It may make no attempt to return to or even towards its original position, but simply depart farther and farther away from it. This is instability with a vengeance!

Then again, whatever the type of stability as outlined above, it may be *inherent*, i.e. due to features incorporated in the design of the aircraft, such as dihedral angle, or *automatic*, i.e. due to some movable device such as automatic slots or an automatic pilot controlled by gyroscopes or other devices.

47. Rolling, Pitching and Yawing

The study of the stability of an aeroplane is rather complicated because of the very involved movements which may be experienced. The only simple way to examine the question is to consider each of the three main types of movement *separately*, although in actual practice they may all occur simultaneously. The result is apt to be misleading, but unfortunately we have no alternative in such a complicated problem.

Let us think of *three axes*, fixed relative to the aeroplane (Fig. 66). The *longitudinal axis*, running *fore and aft* through the centre of gravity; the *lateral axis*, running parallel with a line from *wing tip to wing tip* through the centre of gravity; and the *normal axis* through the centre of gravity, at *right-angles to the two other axes*, i.e. it will be vertical when the aircraft is in straight and level flight. Remember that these axes are fixed *relative to the aircraft*, and that they move with it. For instance, the normal axis is, as stated, vertical in straight and level flight; but it will not be vertical if the nose drops or one wing drops—it will tilt with the aeroplane.

Motion round the lateral axis is called *pitching*. The control or stability, so far as this motion is concerned, is called *longitudinal* control or stability, because it is the longitudinal axis which moves, although the motion is round the lateral axis.

Motion round the longitudinal axis is called *rolling*. The corresponding control or stability is called *lateral*.

Motion round the normal axis is called *yawing*, and the control or stability about this axis is termed *directional*.

48. Longitudinal Stability

Imagine an aeroplane which is flying in normal horizontal flight to be suddenly disturbed in such a way that its nose is raised. If it is to be longitudinally stable, the original forces must change, or new forces must act upon it, in such a way as to restore it to its original attitude.

Let us consider what will happen. Owing to its inertia, the aeroplane will tend to continue, for a short time at any rate, on its original path of flight in spite of the fact that its nose has been tilted upwards. If, however, we were to allow it to remain in its new attitude, it would probably start to climb. If it continues on its old path or very near to it, and remains in its new attitude, the angle of attack of both main planes and tail planes will be increased, i.e. they will both strike the air at a larger angle. On the main planes this will mean an increase in lift; but we must also remember that most wings are in themselves unstable, i.e. not only will the lift increase,

LONGITUDINAL STABILITY

but its line of action will move forward and thus tend to make matters worse rather than better. Thus it is to the tail plane, not the main planes, that we must look for help in obtaining stability (Fig. 67). The increase in lift on the tail plane, or the creation of an upward lift if there had originally been no force, will tend to restore the aircraft to its old attitude.

It is clear, then, that the longitudinal stability of an aeroplane will depend upon the extent to which the correcting turning

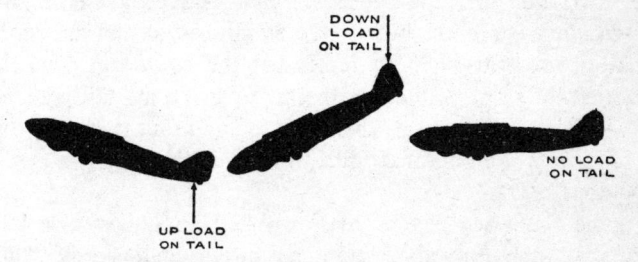

Fig. 67. **The tail plane and longitudinal stability**

effect or moment of the tail can out-balance any upsetting effects on the main planes or other parts of the aeroplane. In deciding this question—and it is a very difficult one—the designer will have to consider the area of the tail plane needed to give enough force, the angle at which it is to be set, and the length of the fuselage which determines its distance from the main planes. He must also take into account the extent of the movement of the centre of pressure on the main planes, the position of the centre of gravity of the aircraft, and various other items. Of these, the position of the centre of gravity is the most important. In order to obtain longitudinal stability the centre of gravity of the whole aircraft must be *well forward*. It is not easy to explain the reason for this in a book of this kind, so you must take my word for it—*a forward position of the centre of gravity helps longitudinal stability*. Let us leave the designer to it—all we can do is to get some idea of the problems which confront him—and after the designer has done his job let all those who load aeroplanes, and even passengers who fly in them, remember that loads must be kept well forward.

It might be mentioned that, as a factor influencing longitudinal stability, there is in most aeroplanes a *longitudinal dihedral angle*, i.e. the main planes are set at a greater angle to the horizontal than the tail plane. This is not an essential for longitudinal stability, but it is certainly usual. In this connection, too, we ought to remember the *downwash from the main planes*. This will cause the air to strike the tail plane at an angle less than that at which it is set, and thus there may be, *in effect*, a longitudinal dihedral angle although none is apparent.

Some modern designs have exhibited an undesirable tendency to pitch upwards, or tighten a turn at high speeds. This, of course, is a form of longitudinal instability and one method of preventing it has been to change the pressure distribution over the wing by having a *notched* or *saw-tooth leading edge*.

49. Lateral Stability

If an aeroplane rolls, forces must come into play so as to restore it to an even keel.

In normal horizontal flight the lift will be vertically upwards and the weight vertically downwards, the two balancing and maintaining equilibrium. If, however, the aeroplane rolls over into some new position, the lift will become tilted, some part of it acting vertically but some part acting *sideways* towards the lower wing. The weight, on the other hand, will still act vertically downwards. Thus there is nothing to balance the sideways component of the lift force, and the aeroplane will *side-slip* to one side, the side of the lower wing.

LATERAL STABILITY

The side-slip, being a relative motion of the aeroplane through the air, will naturally cause a wind to come from the opposite direction to the side-slip (Fig. 68). This wind will strike the wings and all the side surfaces of the aeroplane. If the wings have a *lateral dihedral angle* it is easy to see that the effect of this wind will be to return the aircraft to an even keel,

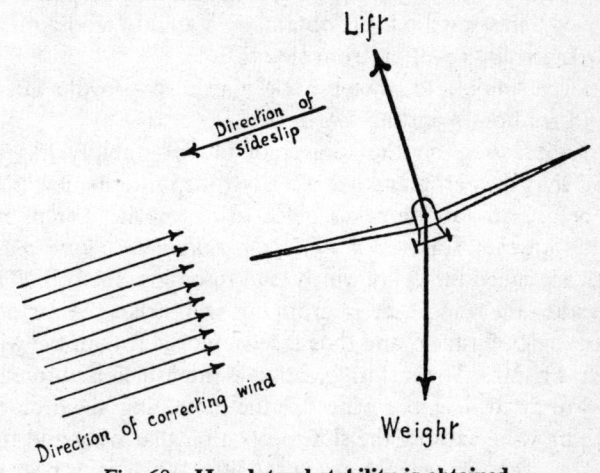

Fig. 68. **How lateral stability is obtained**

and then the side-slip will cease, since the forces will be balanced once more.

This lateral dihedral angle is by far the most common and most effective means of obtaining lateral stability, but, as in the case of longitudinal dihedral angle, it is *not* essential, and other methods are sometimes used.

Some modern types have a measure of negative dihedral, or *anhedral*, on main plane or tail plane or both. This is usually done for very practical reasons such as reducing the length of undercarriage legs, but of course it does have an effect (possibly adverse) on both lateral and directional stability which has to be taken into account in the overall design.

Aeroplanes in which wings are high and the centre of gravity low may be stable without any dihedral angle. Many high-wing monoplanes are examples of this type. The weight, being low, acts like a pendulum when the aeroplane side-slips, the resistance of the wings to their motion through the air holding them back and thus supporting the pendulum. As might be imagined, this method of obtaining stability tends rather towards a rolling motion from side to side.

A large amount of sweep-back may also provide lateral stability without resorting to dihedral.

Our discussion on the subject of lateral stability has assumed that the aeroplane has already rolled into its new position before any action takes place to restore it. Let us not forget, however, that *even while the rolling is taking place*, forces are called into play which tend to oppose the roll. This is because the wing that is dropping will strike the air at a larger angle of attack, and thus receive more lift than the wing which is rising. The restoring action is intensified if automatic slots are fitted near the wing tip, the increasing angle on the dropping wing causing the slot to open on that wing and thus cause an even further increase in the lift. When at or near the the stalling angle the increased angle of the falling wing (if not fitted with slots) may result in a *decrease* in lift—and that is the beginning of a *spin* of which we shall say more later.

50. Directional Stability

Directional stability depends on the fact that, if an aeroplane is suddenly deflected so that it *points in a new direction*, it will temporarily, owing to its inertia, tend to continue to *move in the old direction*. During this crabwise motion it will expose all its side, or keel surfaces, to the air flow. The pressures thus created *on those side surfaces in front of the centre of gravity* will tend to turn it farther off its course, whereas the pressures

DIRECTIONAL STABILITY

created *on the side surfaces behind the centre of gravity* will tend to rotate it back into its original attitude. It is all a question of which influence wins.

In deciding which will win we are concerned not only with the areas of the various surfaces and of the pressures upon them, but also with the *distance* of each surface from the centre of gravity (which may be considered as the turning axis of the aircraft). It is only necessary to look at a view of the side elevation of an aeroplane to see what surfaces will have most effect on directional stability. First, the fin and rudder. These have both large areas and long leverage (Plate 21); in fact, the fin is the determining factor, and the designer decides its area solely from this point of view. Next the fuselage, especially the rear portion of it; the front portion is often of large area, and being so far forward will have a bad effect.

Wheels and undercarriage will not influence stability much either way for three reasons: first, their side surface is not very great; secondly, they are situated almost on the turning axis and thus have very little leverage or turning effect; thirdly, except for take-off and landing, they are tucked well away. The wings themselves, especially if they have a large dihedral angle, present a considerable side surface to the air, but once again their turning effect is small. Even a propeller must be taken into account: its side area may be small, but it is usually the most forward part of the aeroplane, and its turning effect is definitely against stability; the same applies to the forward portion of engine nacelles.

51. Directional and Lateral

Watch a model aeroplane flying on a "bumpy" day; one of the wings will drop, you will see quite clearly the side-slip (see Section 49) towards the lower wing and the recovery caused by dihedral, but you will also notice that the model goes off its course and regains horizontal flight *in a new direction*, having turned towards the wing which was depressed.

In a real aeroplane, if one of the wings drops owing to a disturbance in the air, the pilot will often give a kick to the rudder in the opposite direction.

These two practical instances—both really illustrating the same point—might, at first sight, appear to point to directional *instability*, but they are actually excellent examples of the results of directional *stability*. Lateral stability depends for its effect on side-slipping, *but so does directional stability*. If an aeroplane is to be directionally stable, it must turn *into* a side-slip, and thus the side-slip necessary for lateral stability causes it to turn into the side-slip, i.e. towards the lower wing. In the case of the model, it actually goes off its course—in the case of the real aeroplane, the pilot may prevent this from happening by applying opposite rudder. So inseparable are these two motions, rolling and yawing, that some people prefer to group them both under the one heading of lateral stability.

If an aeroplane is given too little lateral stability—in the narrow sense—and too much directional stability (in practical terms, this means too little dihedral angle and too much fin area), the result of dropping a wing may be as follows: the aircraft will side-slip towards the lower wing, and two opposing effects will be brought into play, the dihedral tending to take off the bank, the fin tending to turn the machine and thus increase the bank. If the latter is the stronger of the two tendencies, the rate of turn and the bank will both increase, the nose will fall and the aircraft go into a steep spiral. This is called *spiral instability*. It should not arise in a well-designed aeroplane.

52. Control

However stable an aeroplane may be, the pilot must have the power to *control* it. The proportion of control to stability

CONTROL

may vary considerably in different types. The helmsman of a ship has only directional control; rolling and pitching depend upon the stability which is provided by the designer. But a ship only travels, intentionally at any rate, in two dimensions, whereas an aeroplane travels in three, and must therefore be controlled in its vertical movement as well as directionally. The submarine provides a better comparison, since it has both fore and aft and directional control. But a submarine is more like an airship than an aeroplane, since it has the power to rise and fall by simply altering its *effective weight*. An aeroplane cannot do this, and is dependent on its fore-and-aft control or alteration of engine power for changes of height.

It may be argued—with some justice—that, although directional and fore-and-aft control are obviously necessary, there does not seem to be any more need for *lateral* control in an aeroplane than in a ship, submarine or airship. For commercial work, long-distance flying and so on, aeroplanes without lateral control and relying entirely on their stability could be used, but they would be more slow and clumsy, and therefore more dangerous, in rapid manoeuvres near the ground, taking off, landing and so on. For all sporting and fighting machines a high degree of lateral control is absolutely necessary.

The standard system of controls is by means of control surfaces hinged at the rear of the tail plane (for longitudinal control), along the trailing edges of the wings (for lateral control), and at the rear of the fin (for directional control). A movement of the control surface will cause a force due to the deflection of the air flow. The effectiveness of a control surface will depend chiefly upon its area, the distance from the axis round which it is intended to turn the aeroplane, and the velocity of air over the control surface.

53. Longitudinal Control

Longitudinal control of an aeroplane is nearly always provided by *elevators* attached to the rear of the tail plane. The principle is best illustrated by the old-fashioned system in which the elevators were connected by control wires and levers to the *control column* in the pilot's cockpit. The control is instinctive, i.e. when the column is pushed forward, the elevators are

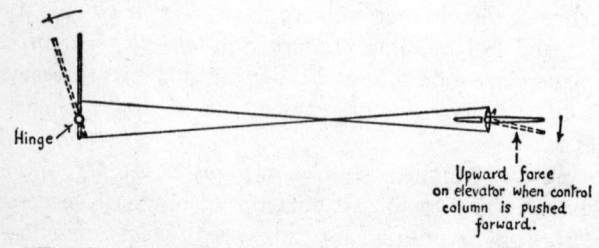

Fig. 69. **Longitudinal control—direction of movements**

lowered and the upward force on the tail is increased, thus causing the nose of the aeroplane to drop (Fig. 69). In order to achieve this result it will be seen that in an ordinary simple control system the wires must be crossed between the control column and the elevators. In modern practice, however, instead of employing two wires which will tend to become slack, causing a certain amount of backlash in the system, more positive controls are nearly always used; these may take the form of a rigid rod serving both to push and to pull the elevators from top or bottom only, or they may rely simply on the torsion of a rod or tube, or the whole control system may be power-operated, hydraulic, pneumatic or electric.

54. Lateral Control

The usual method of obtaining lateral control is by means of *ailerons* hinged at the rear of each main plane near the wing

DIRECTIONAL CONTROL

tips. The ailerons are connected to the control column by a complete system of control wires (Fig. 70), by a rigid system of rods, by torque tubes inside the wings, or again by some power-operated system. This time it is a sideways movement of the control column which moves the ailerons and does so in such a way that once again the control is instinctive, i.e. if the control column is moved to the left the right-hand ailerons

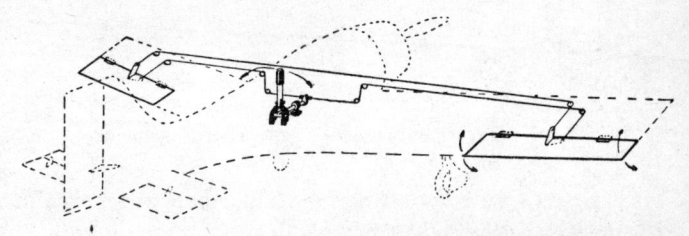

Fig. 70. **Lateral control—general arrangement**

will go down, increasing the lift on the right-hand wings, thus banking the aeroplane to the left; at the same time the left ailerons will have been raised, decreasing the lift on the left wing and thus adding to the effect. Sometimes the control column has no sideways movement, and lateral control is effected by a type of handlebars, or by a wheel very similar to the steering wheel on a car.

55. Directional Control

Directional control is by *rudder*, which has very much the same effect as on a ship. The rudder is connected by wires or rods or by a power-operated system to a rudder bar or rudder pedals on the floor of the cockpit (Fig. 71). In this instance it is not wise to stress the point that the control movement is instinctive, because some people claim that it works the wrong way and should be altered to make it instinctive. If the left foot is pushed forward, the rudder moves to the left (the wires not being crossed) and the aeroplane turns to the left. It all sounds instinctive enough, but it is exactly the *opposite* to

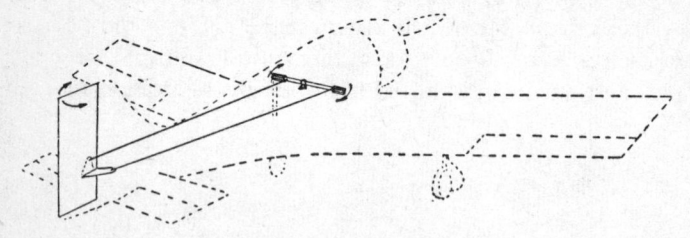

Fig. 71. **Directional control—general arrangement**

what happens on a bicycle when the handlebars are moved in the same way as the rudder bar.

56. Balanced Controls

The actual force which the pilot must exert on the control column, wheel or rudder bar is not very large, but continual movement of the controls during a long flight on a bumpy day may become quite tiring, especially in a large, fast aeroplane. Even when controls are power-operated there are advantages in reducing the loads, and so reducing the necessary strength— and weight—of the various parts of the mechanism. In order to relieve the force which the pilot or the mechanical parts must exert, control surfaces are often partially *balanced*. This may be done by having some portion of the control surface in front of the hinge so that the air striking this front portion will tend to *help* the control surface to move in the required direction. Various methods of balancing control surfaces are shown in Fig. 72.

Great care must be taken *not to overdo this effect*. If we go too far, i.e. if there is too much surface in front of the hinge,

BALANCED CONTROLS

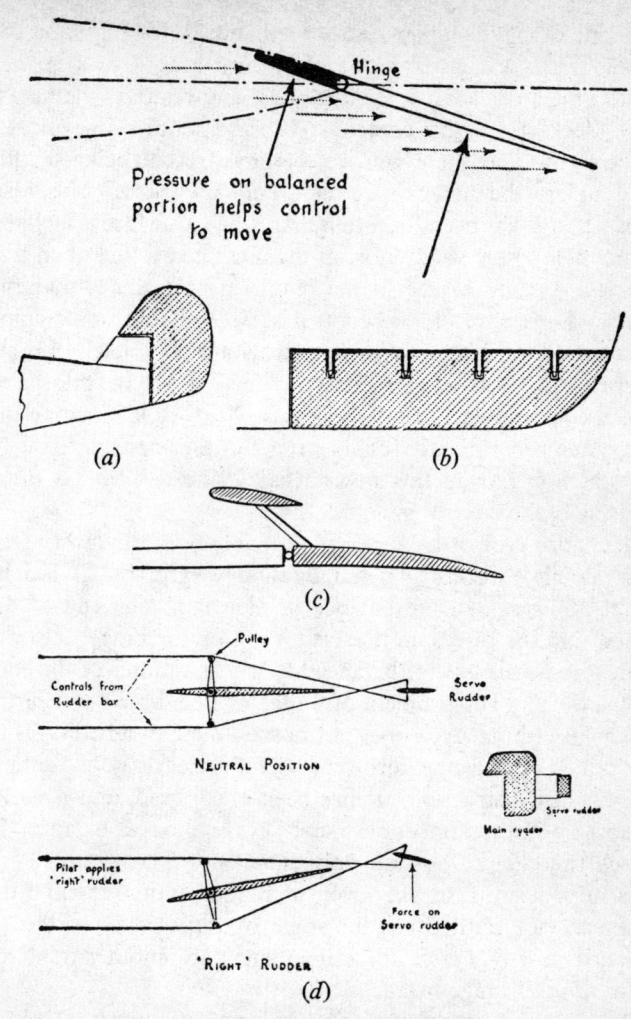

Fig. 72. **Balancing of controls**

(*a*) Horn balance. (*b*) Inset hinge balance. (*c*) Auxiliary aerofoil.
(*d*) Servo rudder system.

the control may be taken out of the hands of the pilot and the control surface tend to move of its own accord to a greater angle. Such a tendency is dangerous, and removes the necessary "feel" from the controls. It is sometimes imagined—incorrectly—that there will not be any overbalancing of this kind unless the hinge is at least half-way back. This idea shows that it has been forgotten that, when a surface is inclined at an angle to the wind, most of the pressure exerted upon it is *near the leading edge*. So the requirement really is that the centre of pressure on the control surface should never come in front of the hinge. When controls are completely power-operated there is—or need be—no "feel" left to the pilot, and this would enable him, without any effort, to manoeuvre the aeroplane too violently for its strength. So important is it to prevent him having this power that artificial "feel" is often built in to the control system.

The horn type of balance has probably been the most used; it is simple to construct and light in weight, but it has its disadvantages. All the balance portion is at one end of the hinge, and the forces on it exert a twist on the hinge; also, at high speeds this type of balance has led to a flutter of the control surface. (The problem of flutter will be discussed later.)

The inset-hinge type does not possess these disadvantages to the same extent as the horn type, but it is clearly not so simple a form of construction. It has been much used, and is easily adaptable to modifications such as graduated balance, in which the balance portion only comes into play gradually as it is most needed, or may even be hidden entirely behind the main surface until a certain angle of the control surface is reached. *Frise ailerons* (described later) are another modified form of inset hinge balance.

The auxiliary aerofoil type of balance is interesting even if only because its merits have never been properly recognized and, in consequence, it has been very little used. By altering

CONTROL TABS

the area of auxiliary aerofoil, any degree of balance can be achieved without changing the design of the main surfaces, and by altering the angular setting of the same auxiliary aerofoil a bias can be put on the controls in one way or another; this is often found useful in correcting flying faults.

57. Control Tabs

Most interesting of all has been the evolution of the control-tab type of balance (Figs. 73 and 74). This seems to have been

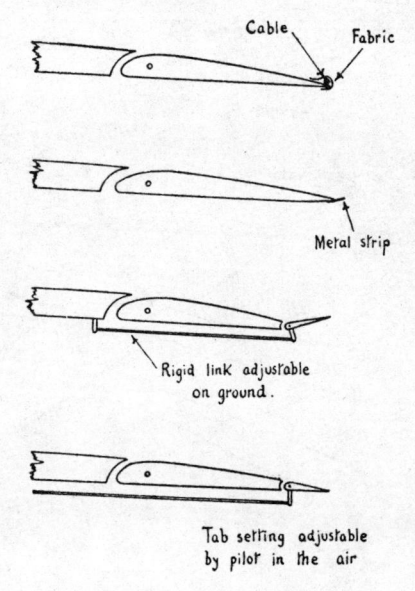

Fig. 73. **Evolution of control tabs**

developed from two different sources; first, from the original Flettner rudder, and secondly, from trimming devices placed on the trailing edge of control surfaces. The Flettner rudder (credit for which is also given to the inventor of the rotor ship already mentioned) was an auxiliary but smaller rudder placed on out-riggers behind the main rudder (Fig. 72d). The auxiliary, or servo rudder, was moved by the pilot in the

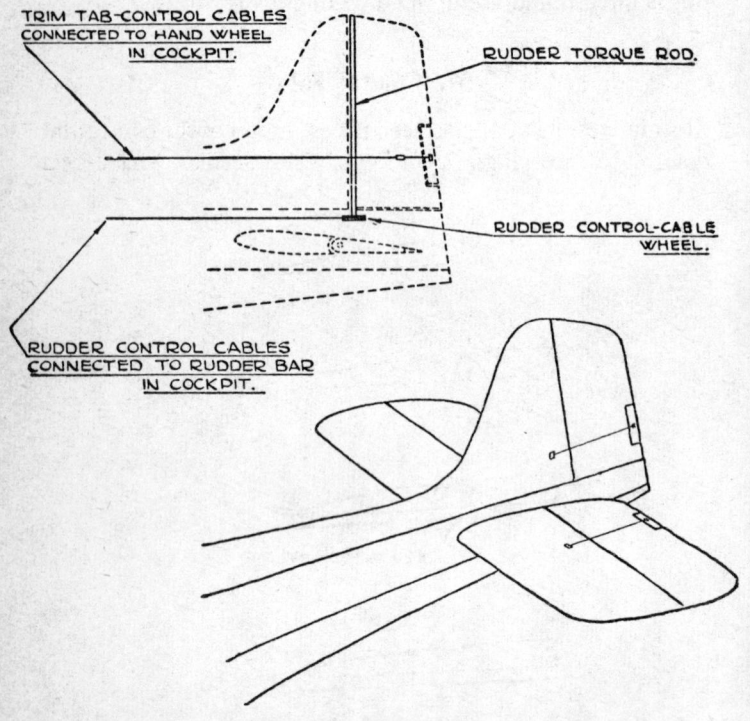

Fig. 74. **Control tabs**

opposite direction to that in which he wished the main rudder to move. The force on the auxiliary rudder, though comparatively small, had considerable leverage about the hinge of the main rudder, and was thus able to turn it in the required direction. Thus the pilot only had to exert a small force which

CONTROL TABS

indirectly moved the large rudder. It sounds rather as though he was getting something for nothing—but this is not really so, any more than one gets something for nothing by using a lever or pulley block and tackle: the effort is reduced, but the distance moved through is greater. The idea of the Flettner rudder was sound enough, and it was adapted for use on many types of aircraft, especially large flying boats.

Apparently originating from entirely different sources came the idea of a trimming strip of flexible metal on the trailing edge of a control surface. If the strip was bent in one direction, it exerted a force which, owing to its leverage, tried to turn the control surface in the other direction; in other words, it acted just like a Flettner rudder. Even before the trimming strip was used, riggers had devised the idea of laying a piece of cable along the upper trailing edge of a Frise aileron, fairing it over with tape and thus deflecting the air upwards, and causing a downward force on the aileron which would cure that wing from flying low. The flexible strip was, of course, not a "balance" at all, but merely a "bias" on the control to cure flying faults. It was but a further step, however, to hinge this strip and connect it to the main aerofoil in such a way that, when the control surface moved, the strip—or *tab*—moved in the opposite direction. Thus we had balance, and, by adjusting on the ground, bias if required.

The final step was to make the tab adjustable by the pilot while flying, so giving both balance and adjustable bias. This final form is both a *control balance tab* and a *trimming tab*. There are not many types of aircraft in which the device is used to its full advantage, but in one form or another it is being utilized in most modern designs. It is simple, light in weight, and effective. Its sole disadvantage seems to be that it may sometimes prove rather too effective, but even this possible defect can be eliminated if a *spring tab* is used. In this device a spring is inserted between the tab and the main control system in such a way that the movement of the tab it reduced at high speeds. Control tabs are easily adaptable so any of the control surfaces, and by avoiding the necessity of the adjustable tail plane (already mentioned in a previous paragraph) they actually reduce complication—a very rare virtue in an aeronautical invention!

58. Control at Low Speeds

Aeroplanes have a striking peculiarity when compared with other means of transport—*the slower they fly, the more dangerous do they become*. Later on this will be explained in more detail; at the moment we are only concerned with *one* aspect of the danger of flying slowly, namely the inefficiency of the control.

In Section 52 we mentioned that the effectiveness of a control depends upon the velocity of the air over the control surface. Obviously the slower we fly the more sluggish and ineffective will be the control. Unfortunately, however, the problem is even more serious than that, especially so far as lateral control is concerned. When an aileron is depressed, the intention is to increase the lift on that wing. That, however, is not the only effect: *the drag will also be increased*. Now, in order to make a correct turn, say to the left, we must bank or roll the aeroplane by raising the right wing, and for this purpose the right-hand aileron will be lowered. The right wing should now go round on the outside of the turn, travelling farther and faster than the left wing. But the increase in drag, mentioned above, *will tend to hold the right wing back* and thus try to prevent the turn, the effect in some instances being very powerful (Fig. 75). This phenomenon is known as *aileron drag*. The ailerons are in fact acting against the rudder and creating a yawing effect in the wrong direction.

This effect is particularly marked at low speeds, because

CONTROL AT LOW SPEEDS

low speed means that the aeroplane is flying near the stalling angle. If the depression of an aileron causes the outer wing to go beyond the stalling angle, not only will this result in a still further increase in drag on the wrong side, but the stalling

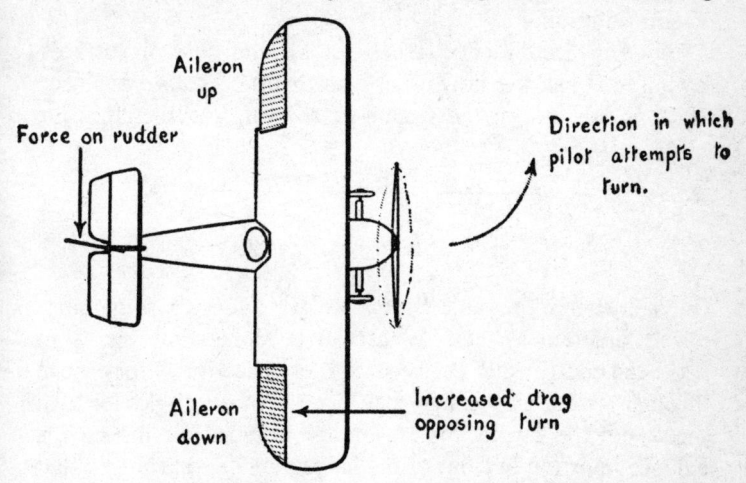

Fig. 75. **Difficulty of a turn at slow speed (large angle of attack)**

of the wing will cause a decrease in lift, which will tend to lower instead of raise this wing and so tend to bank the aeroplane the wrong way.

Much experimental work has been done with a view to solving this problem, or at least mitigating its effect. The need for an aeroplane to fly slowly arises chiefly when near the ground, especially just after taking off and just before landing. In these circumstances even the most modern high-speed aeroplanes must fly comparatively slowly—indeed, paradoxical as it may seem, the more we increase the speed of aeroplanes the more important does the problem of low-speed flight become. This is also when an efficient degree of control is most necessary so that obstacles may be avoided. Further, if any error is made, there is insufficient height in which to recover. Thus it is that the safety of flying is to a great extent involved in this problem, and in the next few paragraphs we shall consider some of the attempts which have been made to reach a solution.

The decrease in effectiveness of all the control surfaces (owing to the slower flow of air over them) presents a problem which is inherent in the system of control, and therefore no

Fig. 76. **Masked balance effect**

real solution is possible unless we try adopting some completely different system. What can be done, however, is to make the control surfaces become more effective as they move through greater angles, and then use the small angles for high speed and the larger angles for low speed. For instance, a balanced portion in front of the hinge may be so arranged that it is hidden behind the main plane, tail plane or fin (according to whether it is an aileron, elevator or rudder) until a certain angle is reached, after which it becomes unmasked, thus proving much more effective (Fig. 76).

Since, however, the main problem is, for all practical purposes, insoluble, attention has been chiefly focused on the particular problem of the aileron control and its tendency to yaw the aeroplane the wrong way.

Attempts to remedy this defect have resulted in the following devices.

(*a*) *Differential ailerons*. These are so arranged (Fig. 77) that the aileron which is raised moves through a greater angle than the one which is depressed. Thus there is less drag and less risk of stalling the outer wing with its depressed aileron, and more drag and less lift on the inner wing.

CONTROL AT LOW SPEEDS 125

It is—to its credit—a simple idea, and one that is certainly a step in the right direction. But it cannot be said to solve the problem.

(*b*) *Frise ailerons*. In these much the same effect is gained by so shaping the ailerons (Fig. 78) that the one which is depressed

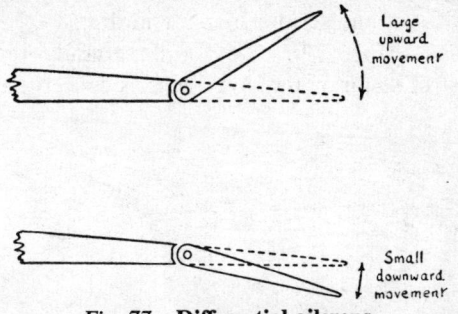

Fig. 77. **Differential ailerons**

maintains a smooth flow of air, thus keeping down the drag and increasing the lift, while that which is raised projects part of the balance portion below the plane, thus increasing drag and decreasing lift.

The advantages and disadvantages are the same as for differential ailerons—the device is simple, cheap and light in weight, but is not sufficiently effective. The two ideas may easily be combined, thus adding to the effect, as shown in Fig. 78.

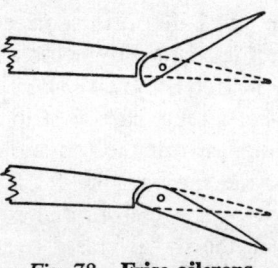

Fig. 78. **Frise ailerons**

(*c*) *Slots*. When discussing aerofoils we discovered that slots increase the lift of a wing, especially at large angles of attack. If we have slots at the leading edge, or slots combined with the ailerons, then the lowering of the aileron on the outer wing will be less likely to cause a stall, and thus the lift will increase without any serious increase in drag.

(*d*) *Spoilers* (Fig. 79). Slots will certainly improve the lateral control so far as the outer wing is concerned. But why

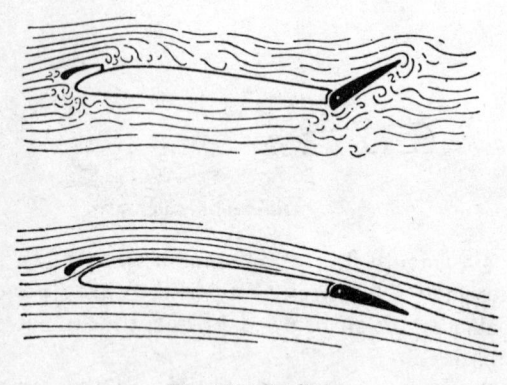

Fig. 79. **Spoilers**
Incorporating also slots and differential Frise ailerons.

not increase drag and decrease lift on the *inner* wing? This method, if somewhat drastic, does at least ensure that the aeroplane will turn and bank in the correct direction; unlike the other methods, it is inclined to be too effective in achieving its aim. The usual method is to fit a *spoiler* just behind the slot. The device consists of a long, narrow strip of metal which can be raised at right angles over the top surface of the aerofoil, increasing the drag and spoiling the lift. It may be worked by a link connecting it to the aileron, and so arranged that the spoiler lies flat when the slot is closed (which it should be at high speeds) and the aileron down. When, however, the aileron goes up, the spoiler is raised, provided that the slot is open. Thus the spoiler does not come into play at high speeds, when it might be dangerously powerful.

It would be very rash to suggest that we have seen the end of inventions conceived with the idea of improving lateral control. It may be that the ailerons will disappear altogether and some spoiler or suction device take their place. Jet engines provide a ready source of compressed air which can be blown over the ailerons, or simply over the wing surface, to increase the lift and prevent stalling; or a small jet itself can be projected upwards or downwards to give a direct rolling effect.

59. Control at High Speeds

But problems of control occur at high speeds too, and although of an entirely different nature, they are demanding more and more consideration as the speeds of aircraft increase.

The first and most obvious trouble is that control surfaces, especially if they are designed to be fully effective at low speeds, are likely to be too effective, too violent in action, or too difficult to move at high speeds—it has not been unknown for pilots to report that they have just not had the physical strength to move the controls of certain aircraft at really high speed. But sometimes the effect of speed may be just the opposite, and the control may become over-sensitive and its effect on the aircraft so violent as to cause structural damage. The only simple answer to these problems lies in graduating the control movement, or the control effect, in such a way that only small movements are needed and occur at high speeds, while the control is geared up, as it were, for low speeds; the masked balance effect mentioned in the previous section was an example of this idea.

A less obvious but perhaps even more important problem arises owing to the *flexibility* of the structure. In saying that, we are embarking on a new part of our subject and we may even be in danger of being lured by its fascination beyond the limits of what can properly be described as how and why an aeroplane flies. But it would be quite unrealistic not to face the fact that an aeroplane, like every other engineering structure, is flexible and that when a load is applied to it, it gives under the load, bending, twisting or whatever it may be. Bridges bend when a train goes over them, church spires and skyscrapers sway in the wind, so it is only to be expected that aeroplane wings, and other parts, should bend and twist when loads such as lift and drag are applied to them, and no alarm need be felt even when the movements of a wing in flight are actually visible, as they often are. Now, when a control surface is moved the change of load on the structure will cause the structure to bend or twist; this means, of course, that its shape will be altered, and this in turn means that the forces on it due to the airflow will themselves be changed. An aileron provides the best example; suppose an aileron is lowered—with the object of lifting that wing—the centre of pressure will probably move backwards (owing to the increase of lift on the aileron itself). So the wing will twist in such a way as to decrease the angle of attack—this means that the *lift on the wing is decreased*; the aileron has in effect acted like a tab to the flexible wing. But the whole object of lowering the aileron was to increase the lift. So the twist in the wing detracts from the effectiveness of the control, especially at high speeds when the angle of attack of the wing will already be small and when the force on the ailerons will be considerable. The consequent loss of control may be serious as may be judged from the fact that it has even been known for the control to be reversed, the lowering of an aileron causing a wing to drop instead of being raised.

But it does not end at that. While it is inevitable that the structure should be to a small extent flexible, it is essential

CONTROL AT HIGH SPEEDS

that it should also be *elastic*, which means that it must return to its original size, shape or position when the loads are removed. Now, the combination of flexibility and elasticity, and the consequent effects on the air flow and on the forces, can easily lead to a dangerous vibration called *flutter*. This phenomenon is difficult for the designer to foresee, and the pilot may have little or no warning of its occurrence. The danger arises in that flutter is one of those vibrations which are liable to develop very rapidly from being comparatively small to such alarming proportions that control surfaces, tail units or even wings may be wrenched bodily from their supports. Vibration is rather like stability; the oscillation may be damped out and gradually disappear, it may remain persistently at the same magnitude, a sort of neutral stability, or it may become rapidly worse, and this last kind is the type that is unstable and dangerous. It is very complicated, and if we cannot understand much about it we can at least console ourselves (except when flying!) with the thought that no one else understands much more than we do. One thing, at least, we can say with a fair degree of certainty—that in order to prevent it the most promising method seems to be to *stiffen up* the whole airframe; in fact to adopt a more rigid kind of structure than has been the practice in the past.

One particular kind of flutter may be set up by the control surfaces, especially the ailerons, and the danger of this may be lessened by fitting accurately calculated balance masses in front of the hinge of the control surface (Fig. 80). This is called *mass balancing*. The mass used is sometimes not noticeable because it is inside that part of the surface which lies in front of the hinge for the purpose of providing the ordinary type of balance. Of course, the weight of that portion in front of the hinge does, in any case, provide a certain degree of mass balance, and so it is not easy to distinguish one from the other, but the ideas are entirely different. The purpose of balance of control surfaces—aerodynamic balance, if we must use that long word—is to make the control easier to move; whereas mass balance is to prevent flutter of the control surface.

The constant changes of the forces on the parts of the structure of the aeroplane, and the consequent stresses and strains and movements one way and then the other, whether or not they result in flutter, may after many reversals of load

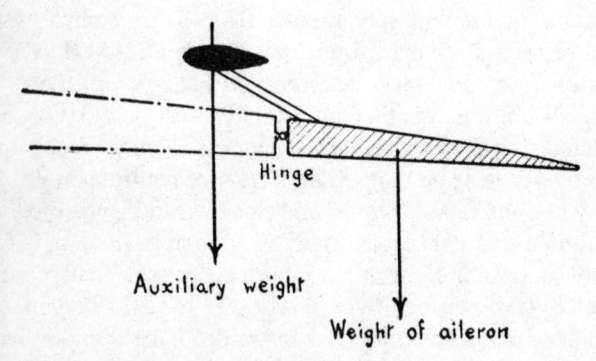

Fig. 80. **Mass balance**

(sometimes running into millions), produce another phenomenon called *fatigue*, resulting in the material after a long life losing much of its strength.

Just as the structure should be rigid to prevent flutter, so the working of the control surfaces should be as positive and definite as possible. That is why rigid rods have tended to replace control wires, and why both elevators must be rigidly connected together. But the flutter problem has led to an even more radical change, i.e. *irreversible and power-operated controls*. In these there is no *feel* to the control; if an aileron is depressed it will stay there, the force from the air having no effect on it, and the pilot will feel nothing tending to restore the control column to its neutral position. This at first seemed rather alarming, and pilots did not take to it kindly; but the irreversible or power-operated type of control is now generally accepted as the system on large and high-speed aircraft, though usually some feel is introduced artificially so that the pilot *thinks* he is having to push or pull to move the controls—whereas in reality he is only pushing against a spring or some other device.

60. Level Flight—The Speed Range

But we were led to this long discussion on stability and control during a consideration of the forces acting on an aeroplane in straight and level flight, and we must now return to that subject and consider the whole range of speed over which a given aeroplane can fly straight and level.

There are sometimes good reasons for wanting to fly fast, perhaps even as fast as possible, but it is a most extravagant kind of flying. Less often there are good reasons for wanting to fly as slowly as possible; this is usually for landing or some such purpose and doesn't continue for long—continued slow flying is also very extravagant, in fact even more so than fast flying, because with all the high fuel consumption we don't get anywhere very quickly!

The range of speeds over which a particular aeroplane can fly in straight and level flight is a most important feature of that aeroplane, and depends mainly on efficiency of design and engine power. There is a story, from the early days of light aeroplanes with very small engine power, of a competition to test speed range in which a certain aircraft when flying as slowly as possible flew faster than when it was flying as fast as possible—in other words, its speed range was for all practical purposes nil; it could only fly at one speed. But we have made a lot of progress since those days, and maximum speeds of ten times the minimum speed, i.e. speed ranges of 10 to 1, have now been achieved.

Let us consider how it is that an aeroplane can keep in level flight at different speeds, and why its speed range is limited. The determining factor is that the lift must equal the weight. Now (other things being equal) lift increases with the square of the speed (Section 21)—but (other things being equal) lift also increases with the angle of attack (Section 19). It is these two salient facts which give the clue to the speed range. In order to fly as fast as possible the pilot decreases the angle of attack and increases the engine power (because the drag increases with speed) until, even with full power, he can no longer maintain level flight; he has then reached the *maximum speed of level flight*. In order to fly as slowly as possible the pilot increases the angle of attack, and rather surprisingly again increases the engine power (because the drag increases with angle of attack) until, even with full power, he can no longer maintain level flight; he has then reached the *minimum speed of level flight*. The difference between the maximum and minimum speeds, or, as it is sometimes expressed, the ratio of maximum to minimum speed, is the *speed range*.

Ever since the early days of flying, since the days of the aeroplane which could only maintain level flight at one speed, or over a very small speed range, we have tried to increase the speed range in both directions. But it is interesting to note that until recently the extension has been almost entirely at the high-speed end of the range, and this at the expense of the low-speed end—in other words, *as the maximum speeds of aeroplanes have increased so too have the minimum speeds*. The reasons for this are simple—the increase in maximum speeds has been achieved in the main by increased engine power together with improved shape and decreased size and wing area of the aeroplane in relation to its weight; factors which have militated against a reduction in minimum speed. Until we

LEVEL FLIGHT—THE SPEED RANGE

approached the speed of sound (see later sections), nothing seemed to stand in the way of increased speed (and in retrospect even the sound barrier does not seem to have been a very formidable obstacle), but at the low-speed end of the range it has been a very different story, and even now, when we can claim to have achieved *vertical take-off and landing* for certain rather freak types of aircraft, we are still a very long way from an aircraft which can fly really fast and really slowly—let alone stop in the air.

Usually, however, we just want to fly at some reasonable speed, and provided we can do so with comfort, we might as well be economical about it. To some extent the problem is the same as when driving a car. Driving too slowly is uneconomical—and often unsafe, though for quite a different reason from aeroplanes. Driving too fast is certainly uneconomical—and usually unsafe, which does not apply to aeroplanes. So we cruise at some reasonable economical speed, getting as far as possible for our petrol and for our money. But in flying there is an additional and most important aspect of the problem; we cannot stop just when we wish to fill up. We can only stop at a suitable airfield, and even so, we must descend, land, taxy, fill up, taxy again, take-off and climb before we are once again on our way. This takes time and incidentally uses petrol. Sometimes, too, such airfields are few and far between, and in wartime over hostile territory are denied to us altogether. So we must take our fuel for the whole journey with us on the aeroplane (that is unless there are facilities for refuelling during flight which are not very common even in these days); in peace time it may be just a single journey from A to B—though we must allow for diversions owing to bad weather—but in wartime we may have to fly out to the target and back without a chance of refuelling. So it is not just a question of being economical so far as money is concerned, it is also a question of getting there or getting back at all, and, furthermore, we want to do it on the least possible amount of fuel, because more fuel means more weight to carry, more weight means more lift is necessary, more drag, more power—and less economical flying—a vicious circle if ever there was one!

61. Economical Flying

But what will be our most economical cruising speed? *Each gallon of fuel, if used as economically as possible by the engine, will enable us to do a certain amount of work*, that is to say will give us so much *force × distance*—we are getting precariously near to formulae, but never mind. Note that we shall only get this force × distance if we use the engine economically and that is where engine handling comes in—a subject in itself. Now, if the force × distance that we can get is fixed, *we shall clearly get the greatest distance* if we push with the least force, that is to say *if we fly with the least possible thrust, the least possible drag*. When an aeroplane flies very fast there is a lot of drag—form drag and skin friction—didn't we say that these forms of drag go up with the square of the speed? So we must not fly too fast. But when an aeroplane flies too slowly there is also a lot of drag; this time it is the induced drag which goes up, because slow speed means large angle of attack, a big difference between the decrease in pressure on top of the wing and increase underneath, large deflections of the airflow and violent wing-tip vortices (you may remember that we made a confession about this when we first mentioned induced drag in Section 26, and had to explain that induced drag was a glaring exception to the rule that drag goes up with the square of the speed). So some kinds of drag go up when we fly fast, another kind when we fly slowly; what is the happy mean? Whether we fly fast or slowly *the lift must be kept equal to the weight, so we will have least drag when we fly*

ECONOMICAL FLYING

with the best lift/drag ratio, which means presenting the aeroplane to the air in a certain attitude, and this in turn can be translated for the pilot into terms of air speed. This is *the air speed for maximum range*—it will enable us to fly the greatest possible distance on a given quantity of fuel *if there is no wind*. The wind is an unfortunate but very important complication; its importance can easily be seen by a little exaggeration. Suppose we encounter a head wind equal to the best air speed (for range) of the particular aeroplane; if we fly at this speed we shall get nowhere and it will obviously pay us to fly faster, use more fuel, and get somewhere. Even if we are flying from A to B and back to A, and on either leg this head wind is against us we shall never get there and back. So in practice we must allow for wind, by flying faster when against the wind, and more slowly when the wind is with us (reminiscences of our first considerations about air speed and ground speed!).

There are times when we are not particularly interested in getting anywhere, but for some curious reason *wish to stay in the air for as long as possible*—perhaps to break a record, or to search for something or for some display or other. This is called *flying for endurance*, and, strange as it may seem, this is not the same problem, nor does it require the same air speed, as flying for range. We got range flying with the least possible drag and we were not concerned *with the rate of fuel consumption; in flying for endurance this, of course, is all important and it depends not on the drag encountered but on the power used*, that is to say on the rate of doing work. By flying more slowly than the speed for maximum range we shall have more drag, but if we choose the correct speed we shall actually use less power because power is *drag × speed*, and the greater drag is more than compensated by the lower speed. So—and this applies to all aeroplanes—the speed that gives best endurance is lower than the speed that gives maximum range. In practice, both speeds are rather low for comfortable flying, and unless it is a matter of life and death recommended speeds are usually rather higher.

In order to obtain economical flying there are, of course, other things to be decided besides the speed; the chief of these are the load to be carried and the height at which to fly. The first is easy and has, in effect, already been answered. Load means lift, lift means drag, and drag means power; so whether we want maximum range or maximum endurance we should carry the least possible load, not only of fuel but of everything else. The problem of height is not quite so simple. From the aeroplane's point of view it makes no difference at what height we fly in order to get maximum range, and the actual height will be decided by the efficiency of the engine and of the propulsive system at different heights—and, of course, the wind. From the aeroplane's point of view we shall use least power and therefore get maximum endurance if we fly low, at sea level if possible—but again other considerations may influence the decision.

When using propeller propulsion it is, in general, true to say that it is the aeroplane's point of view that decides the issue—i.e. the speed and height at which to fly. The efficiencies of the engine and propeller are influences, but they are not the main determining factor. Not so with jet propulsion. *The faster we fly*—within reason—*the more efficient does this system become*, and incidentally we don't use much more fuel. *The higher, too*—within reason—*the better*. So important are these factors that it pays us to disregard to some extent the claims of the aeroplane, which are the same for any kind of propulsion, and to step up the speeds for economical flying whether for range or endurance, and to fly higher. With rockets, both speeds and heights need stepping up even further.

These are great and growing problems, and clearly of immense importance in all commercial flying and in nearly

STALLING

all war flying; in fact, it is only in record breaking or extreme emergency that they can be neglected. We cannot pretend to have covered the problem in full—it would take a book to do so—but we hope that we have said sufficient to create an interest in this aspect of flying.

62. Flying at Low Speeds

But let us return to one of the most interesting problems of flight, that of flying at really low speeds. In the last section it was said that the economical speeds of flight are rather low for comfortable flying. But the most important fact about flying, as distinct from other means of transport, is that at some speed not much below these economical speeds *flight becomes not only uncomfortable—but impossible*. Why? Well, if we have followed the arguments so far, the answer is simple.

We approach the minimum speed by increasing the angle of attack, and the attitude of the aeroplane to the air, so as to compensate for the loss of lift due to loss of speed—at the same time we increase the engine power to overcome the increased drag (mostly induced drag). But there is a limit to what we can do in this direction.

63. Stalling

In Section 19 we examined the question so far as the *aerofoil* was concerned. When a certain angle was reached, eddies were formed, drag increased tremendously, and lift decreased. All this happened even if the speed of the air flow over the wing was kept constant, e.g. in a wind tunnel. Now, it is not quite the same story when considering an *aeroplane* in actual flight. If, when flying in normal flight, the angle of attack is gradually increased, the speed will also decrease, there being a definite speed corresponding to each angle of attack. Therefore, when the stalling angle of the aerofoil is reached, the aeroplane will be flying at a speed called the *stalling speed* of that aeroplane. *If the angle of attack is still further increased, the speed will continue to fall.* Thus the lift will decrease rapidly, and for two reasons. First, because the angle of attack is past the stalling angle, and secondly because the speed is decreasing. Since the lift only just equalled the weight at the stalling speed, it will obviously fall below it when the speed is less than the stalling speed, and therefore the aeroplane will start to fall, usually dropping its nose rapidly and going into a dive until it regains flying speed. It is in this dropping of the nose, and diving before speed and control are regained, that the danger lies. Considerable height may be lost, and if the original height was not sufficient, the aeroplane will strike the ground.

Now, it is one thing to stall when the wheels are just skimming the blades of grass or the runway, which, in effect, is what we do in landing, but it is quite another thing to stall when you are 10 ft, 50 ft, or even 100 ft off the ground—and, unfortunately, it is easy for a pupil to do this just after taking off, or just before landing. Stalling, in some form or other, is probably the greatest danger of flight, and directly or indirectly it has been the cause of the majority of aeroplane accidents.

For it is not only when in level flight that accidental stalling may occur; when climbing or gliding, and more especially in a gliding or climbing turn, it is even more likely, because it is very difficult to judge the angle of attack when the aircraft is in these attitudes and one has to rely entirely on instinct or an air-speed indicator, and the speed is liable to decrease rapidly because of the absence of thrust when gliding, and the steep attitude when climbing. Furthermore, when considering manoeuvres, we shall find that in some circumstances the stalling speed may be considerably higher than in normal flight.

In some small types of aircraft automatic slots have been used to prevent accidental stalling. When the aeroplane

LANDING

approaches the stalling angle (or stalling speed), the slot opens and the stall is postponed until a much greater angle is reached. If the pilot persists in tempting providence by approaching this new angle, he may get what he asks for—but he has been warned.

What is really needed as a stall warning is an instrument to show the pilot *at what angle of attack he is flying*—many such instruments have been devised but none has proved really suitable to meet all the conditions in which stalling may occur.

But our main interest in the fact that an aeroplane has a minimum speed of flight lies in its influence on landing and taking off.

64. Landing

The essential problem of landing is that of being able to fly slowly—and that, as we have said, is one of the main problems of flight.

The art of landing is to transfer the aeroplane from the medium in which it has been flying—namely the air—as gently as possible on to the ground. In order to approach the ground the aeroplane must have both a forward and a downward velocity. The forward velocity—*relative to the ground*—is reduced by landing head to wind (Fig. 81), though, with the

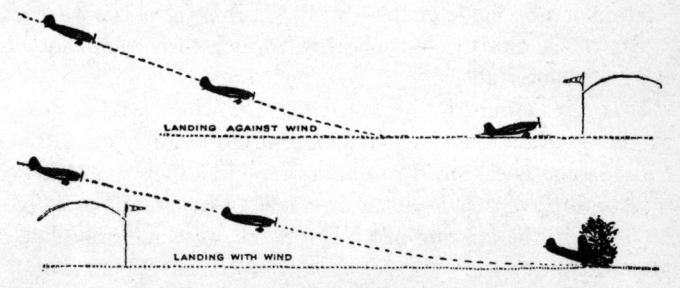

Fig. 81. **Landing with and against the wind**

modern prevalence of runways, it may often be necessary to land with the wind on one side, and this involves a special technique. When very close to the ground the forward velocity —*relative to the air*—is reduced to the minimum at which flight is possible. This is effected by increasing the angle of attack (by raising the elevators), so that, as the speed falls, the lift is still kept equal to the weight by the increase of angle. This process cannot go on for ever (as explained in Section 63), because eventually we shall reach a stage when an increase in angle will, *in itself*, cause a decrease in lift, quite apart from the falling-off in lift due to the decreasing speed. Ordinary flight at any lower speed is impossible for that particular aeroplane.

The slowest and most effective landing will be made if the aeroplane touches the ground just exactly when this condition of flight has been reached. It can no longer fly, its wheels are just about to touch the ground, and thus it subsides gently on to the ground. In order to achieve this state of affairs it must be possible to incline the aeroplane so that its wings are striking the air at an angle of at least 15° before the tail wheel touches the ground. For this reason it will be found that when an aeroplane (not one with a tricycle or nose-wheel undercarriage) is resting on the ground its wings are inclined at about 15° to the horizontal, and if this is so, it means that a three-point landing can be made at the lowest possible speed, i.e. the two main wheels and the tail wheel will touch the ground just as the aeroplane stalls.

It is interesting to note that when the wings are very close to the ground there is a slight, but noticeable, cushioning effect, sometimes called *ground effect*—in other words there is just a little of the air-cushion vehicle, or hovercraft principle, involved in the landing of an aeroplane, especially one like a glider in which the wings approach very close to the ground.

Slots raise a problem in this connection. When slots are

LANDING

used, the aeroplane can fly at a larger angle of attack, and therefore more slowly, without stalling. Thus the landing speed may be reduced, *provided such an angle of attack can be reached before the tail wheel touches the ground*. This means a very high undercarriage, which, of course, will add to the drag of the aeroplane when it is lowered, and which will be difficult to make retractable. An alternative method would be to have the wings adjustable so that their angle relative to the fuselage can be altered during flight just as some tail planes are. Unless such a device is used, slots cannot be employed to their full advantage in so far as reducing landing speed is concerned.

It is not *always* necessary to land as slowly as possible, and good landings may be made on smooth ground at speeds much higher than the stalling speed. The aeroplane will, however, land with its tail up, and the length of run after landing will be much increased. There will also be some danger of striking bumps on the ground which will cause the aircraft—which still has flying speed—to leave the ground again. If the pilot is careful, this may not matter very much; but on the other hand, it may lead to dangerous bouncing. This is where the *nose-wheel or tricycle undercarriage* is interesting. With this device the centre of gravity of the aeroplane is in front of the main wheels, and the aeroplane is prevented from going on to its nose by an extra wheel farther forward. In this way, two great advantages are achieved. In the first place, the aircraft becomes directionally stable when taxying (see Section 72). Secondly, the danger of bouncing is lessened because even if the aeroplane lands at high speed it will pitch on to its front wheel, the angle of attack will be reduced, the lift reduced, and it is less likely to bounce into the air again.

If aeroplanes are to be in ordinary use among ordinary people (a state of affairs that has not been achieved after more than sixty years of power-driven flight), *they must be capable of landing in a small space*. In order to obtain this ideal they must, first of all, be able to land slowly, which is the same thing as saying to fly slowly, *but that is not all*. They must be able to *approach the ground at a steep angle* so that they can avoid obstacles, such as buildings, trees, telegraph wires and such-like on the boundaries of the landing-ground and also touch the ground as close as possible to the near boundary (more will be said about this under the heading of gliding). Finally, *they must be able to pull up quickly after landing*.

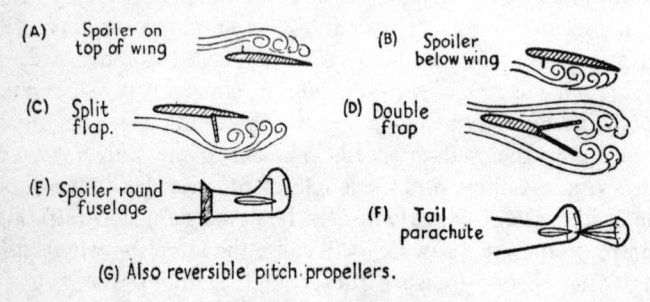

(A) Spoiler on top of wing
(B) Spoiler below wing
(C) Split flap.
(D) Double flap
(E) Spoiler round fuselage
(F) Tail parachute
(G) Also reversible pitch propellers.

Fig. 82. **Air brakes**

These important points have been sadly neglected until recently, and it was at one time considered a silly idea to fit an aeroplane—the fastest means of mechanical transport—with brakes. But then, retractable undercarriages, streamlining and lots of other things which we know to be sensible today were at one time considered silly. The run after landing can be reduced by wheel brakes and air brakes. An air brake is any means of increasing the air resistance, and the wings themselves, when inclined at a large angle, form a very efficient air brake. Air brakes cannot, of course, reduce the actual landing speed, i.e. the speed of flight *just before* landing, but they can reduce the run *after* landing. Air brakes are becoming important in flight too, and Fig. 82 illustrates some of the ideas that have been tried (see also Plates 60 and 61).

Another method, and a very effective one, is to reverse the thrust of the jet.

65. Reduction of Landing Speed

Can we expect aeroplanes of the future to land more slowly? The answer is probably "yes." We must remember, however, that we are also trying to increase speed in the other direction, and the more we increase the maximum speed, the more difficult does it become to reduce the minimum speed. We have really done very well in keeping the minimum speed of the average aeroplane more or less the same while maximum speeds have been steadily raised.

Let us consider what we must do to reduce landing speed.

The aeroplane is in flight just before it touches the ground. To maintain flight, the lift of the wings must be equal to the weight of the aeroplane. The lift of the wings depends on:

(a) *The shape of the aerofoil section.* (More top camber, more lift.)
(b) *The angle of attack.* (More angle, more lift—but only up to the stalling angle.)
(c) *The air density.* (Greater density, greater lift.)
(d) *The wing area.* (Greater area, greater lift.)
(e) *The velocity.* (Greater velocity, much greater lift.)

All these go to make up the lift; so if the last one, the velocity, is to be as small as possible, all the others must be as great as possible.

Thus we shall reach the lowest landing speeds, taking into account the five points mentioned above, with (a) *a high-lift, big-cambered, thick aerofoil section.* But this will spoil the maximum speed. So what we really need is *variable camber* or *suitable flaps*—the big camber to give us low speed, the small camber high speed.

In recent years many new flaps and slotted flap devices have been invented, with varying success, with the idea of decreasing landing speed. Some of these, at the same time, act as air brakes and thus serve the useful purposes of steepening the angle of glide and decreasing the run after landing, the two

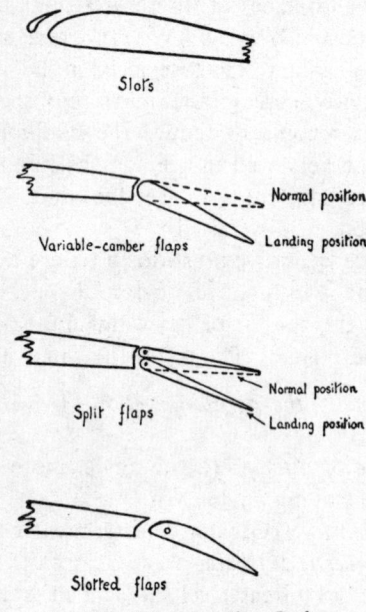

Fig. 83. **Slow-landing devices**

special instances where large drag is an advantage. The illustrations (Fig. 83 and Plates 42, 60 and 61) show some of these devices, and it will be noticed that the slot and the flap are sometimes combined in the form of a slotted flap. Nearly all flaps and slots, and indeed the wing itself, can be made more effective as lifting devices at low speeds by blowing air over them, and the air compressed in jet engines provides an effective means of doing this.

(b) *Angle of Attack*. About 15° is the maximum for an ordinary aerofoil—improvement is made by *slots* or other artificial means of increasing stalling angle and thus prolonging the increase in lift. Unfortunately, however, very large angles of attack are awkward for landing, since they necessitate a high undercarriage.

(c) *Air density*. This is outside our control. Landing speeds are noticeably lower at sea level than in mountainous country, and in low temperatures than in high. But the air density is an act of God, and we are helpless to alter it.

(d) *Wing area*. Increase the wing area, and we shall decrease the landing speed. That is not quite so easy as it sounds. It does, however, account for the ridiculously large wing areas which make many aeroplanes look so clumsy, and *which hamper so seriously their maximum speed*. Another point is that if we increase the wing area still more, we shall increase the weight, and so our lift will have to be *still greater* to balance the new increased weight, and we shall have gained nothing. So far as maximum speed is concerned, retractable wings which could be drawn in for high speed and spread out for landing would seem to be a promising idea. But such devices mean extra weight and additional complication to an already over-complicated mechanism. However, variable wing area, in some form or other, may yet come into its own; and variable sweep, by swinging the wing, has already been proved to be a practical proposition.

66. Wing Loading

We now appear to have exhausted the possibilities; but that is not so. We have considered means of increasing lift. The lift must be equal to the weight. Can we, perhaps, reduce the *weight*? This, too, will enable us to lower landing speed. Here is a fascinating problem, but unfortunately one that takes us outside our subject. It is a question of materials, of structural design, and there are still further possibilities, e.g. lighter metals, lighter alloys with greater strength, completely new methods of design: there is always hope in these directions, and such possibilities will help us to reduce landing speeds.

We can best couple the ideas of weight and wing area together and talk of *wing loading*, i.e. the weight per square foot of wing area. In actual figures, wing loading may vary from 1 or 2 lb/sq ft on gliders to 7 or 8 lb/sq ft on light aeroplanes, and 20, 30, 40, 50, 100 or more pounds per square foot on high-speed aircraft. As the wing loading goes up, so does the landing speed—other things being equal. An aeroplane may be assisted in its take-off by a catapult or by rockets or by riding into the air on the back of another machine, but the landing cannot be assisted in the same way. There is, however, one consolation. In the normal course of events, i.e. except in case of a forced landing, the aeroplane will have completed its journey before it lands, and therefore most of the fuel will have been used, with a corresponding reduction in weight and thus of landing speed. Possibly, too, mails or bombs may have been dropped, or other weights released.

67. S.T.O.L. and V.T.O.L.

With ever increasing maximum speeds—and wing loadings—the battle for lower landing speeds might have been given up as lost if the *Cierva autogiro* (Plate 38) had not appeared in the early 1930s. Now that we have become accustomed to seeing things with rotating wings flying about the sky it is not easy to realize that it was a long time before they became practical propositions; and that they were first received with scepticism. For, as one aeronautical expert put it, "Aeroplanes were surely never meant to look like that!"

But, the autogyro accomplished in one fell swoop greater

reductions in landing speed than had been obtained in years of patient experiment along orthodox lines. The reason for this is quite simple; i.e. that, although the forward speed of the *aircraft* may be very low, the speed of the rotating *wings*, relative to the air, may be such as to provide sufficient lift.

In a true *autogyro*[1] the wings are not driven by an engine; they are free to rotate; hence of course the name "autogyro," which, for this reason, the author prefers to the official name of *gyroplane*. The aircraft—officially a rotorcraft, not an aeroplane—is propelled through the air in the usual way by engine and propeller. By inclining the axis of rotation slightly backwards the wings are caused to rotate by the motion through the air—and the consequent drag on the wings which differs between the wing that is going forwards and the one that is going backwards (relative to the motion of the aircraft). The rotation of the wings in turn gives the lift. It sounds remarkably as though we were getting something for nothing—but we are not (we never do!)—the power is all coming, however indirectly, from the engine and propeller.

But notice that the wings of the autogyro only rotate, and therefore obtain lift, *due to the motion through the air*; in other words, forward speed is necessary, as in other types of aircraft that are heavier than air; it is only a question of degree. The autogyro was, in fact, a remarkable example, a pioneer of what in modern terminology would be called an S.T.O.L., a *Slow Take-off and Landing* type of aircraft.

Even in conditions of no wind the run required by the autogyro, both for take-off and landing, was remarkably small, and by various tricks—such as speeding up the rotating wings before take-off—could be reduced to nothing.

Experiments were also made with another type of rotorcraft, the *cyclogyro*, in which the wings rotate about a horizontal

---

[1] The spelling "Autogiro" was a trade name reserved for the original Cierva type.

axis like paddle wheels—but this type has never shown much prospect of success; nor has the flapping-wing *ornithopter*, which is reminiscent of man's earliest attempts to fly.

But by this time the true *helicopter* had arrived (Plate 39). Now, the helicopter, in its crude principle, is easier to understand than an autogyro, easier perhaps than a conventional aeroplane; the idea must be nearly as old as the screw itself, and flying models of one type or another have been made by schoolboys for many generations. Why, then, did it take so long for it to become a practical proposition? The answer is the same as the one that provides the clue to the delay in producing a practical aeroplane of the conventional type, i.e. *the difficulty of getting a sufficient ratio of power to weight out of an engine*, but with the helicopter the problem was even greater because it was necessary to produce *a thrust equal to the weight of the aircraft*, whereas in the conventional aeroplane, as I hope we understand by now, the thrust (equalling the drag) need be only a fraction of the weight (equalling the lift).

The crude principle of the helicopter may be simple, but in fact there were some difficult problems to solve, in addition to that of power/weight ratio, before it could reach even its present stage of development. To every action there is an equal and opposite reaction, and by driving the wings round in one direction we make the aeroplane try to rotate in the opposite direction (this, as we shall mention later, tends to happen in the conventional aeroplane, but to a lesser degree, and whereas in the conventional aeroplane it is a tendency to to roll, in the helicopter it is a tendency to yaw); this does not happen in the autogyro because the wings are not power driven. Another difficulty is that the wing travelling in the same direction as the aircraft has a higher speed (relative to the air) than the wing travelling in the opposite direction, and so has more lift, and more drag—or would have if steps were not taken to prevent it. This happens in both the helicopter and the autogyro. A remarkable degree of ingenuity has been displayed in the solving of these and other problems, but unfortunately the solutions have involved complications which have added to the complication already present of having rotating wings as opposed to fixed wings. Both these main problems can be solved by having two sets of wings rotating in opposite directions; this was tried in both autogyros and helicopters, and in some instances is still used. In general, however, the tendency to yaw is prevented by having a small propeller at the tail (Plate 39), which also serves as a rudder, and the tendency to roll owing to unequal lifts is prevented by automatic changes in angle of attack and dihedral of the wings as they rotate—giving them in effect a flapping motion.

The helicopter, with all its faults, is in modern terms a true V.T.O.L., i.e. a *Vertical Take-off and Landing* type of aircraft, and as such has great possibilities for the future, particularly from the civilian point of view, for flying between city centres and airfields on the outskirts, and from the military point of view, for taking off and landing in places where there are no airfields. Its shortcomings are in noise, in complication (with its potential unreliability), and in serious limitations as regards high speed. The ingenious combination of autogyro and helicopter as displayed in the Fairey Rotodyne (Plate 40) extended the speed range but unfortunately was not a commercial proposition; or so it seemed at the time. However, as too often happens, some of the principles involved have now resulted in the development in America of a helicopter with a rotor mounted on a fixed axis, surmounted by another smaller rotor acting as a gyroscopic flywheel; eliminating, as it seems, most of the helicopter problems—including that of high speed. But even that is not the last to be heard of helicopter modifications; for some time it has been possible to fold the wings of the rotor backwards for stowage on board ship; but why not do this in the air, and even stow them in the fuselage? Retractable rotors like retractable undercarriages, truly vertical take-off and landing, and all the advantages of the normal aeroplane for cruising and high-speed flight. It can be done; it has been done.

The autogyro and helicopter, in practical form, came as a surprise to many people; and they were further surprised that these aircraft did not die a natural death but, on the contrary, have gone from strength to strength. They are dealt with in much more detail than is possible here in *The Helicopter and How it Flies*, a companion volume in this series.

But the autogyro and helicopter are not the only means of obtaining vertical take-off and landing, and we have already come a long way from the "flying bedsteads"—as the first devices for using jets to provide vertical thrust were rather rudely called (Plate 43). By using separate jet engines to provide vertical and horizontal thrust, or, more economically, by deflecting jets from the vertical to the horizontal, we can not only take off and land vertically, but can overcome one of the great disadvantages of the helicopter, its rather serious limitation as regards high speed in horizontal flight (Plate 44). Aircraft employing such devices have already been proved to be practical propositions, and there is no doubt that further improvement will come with experience—and who knows what completely new ideas may not be forthcoming? (Plate 41).

68. Gliding

Gliding is one of the most delightful and one of the most interesting conditions of flight (Plate 46). It is delightful because as the roar of the engine dies away one loses the noise, the vibration and the battering and eddying of the slipstream. It is like free-wheeling through space: it *is* flying. It is interesting because it is the best test of efficiency, whether of pilot or

GLIDING

aeroplane. The most efficient aeroplanes will glide the farthest —in the hands of the best pilots.

We are considering, in this book, the art of gliding in so far as it concerns the normal power-driven aeroplane; but although it does not always seem to be realized, there is no fundamental difference between the glide of an ordinary aeroplane and that of a sailplane. To the latter, gliding is a whole-time job, a profession; to the former it is a spare-time job, a hobby. But football is no less football because it is only played on a Saturday afternoon; it may become more efficient football, more polished, more perfect if it is played on all the other days of the week, but that is only because it then becomes the most important thing to be considered, instead of being a side-line. So it is with gliding. The sailplane is designed for better gliding than the power-driven aeroplane, because gliding is its job. However, the principles are the same for both.

When the engine is cut off, the thrust—one of the four main forces—will cease to exist. There remain lift, weight and drag. These cannot maintain level flight, since if the aeroplane remained in the same attitude there could be no force to balance the drag, speed would fall off, and lift would be lost and could no longer equal the weight. The aeroplane would commence to fall under the action of gravity—the probable result being a spin. However, looking back to Section 43, we find that the forces will probably be so arranged that, if the thrust does cease, the aeroplane will tend to take up a natural gliding position, the nose pointing slightly downwards. In this way part of the weight takes the place of the missing thrust, speed is maintained, and the forces of lift and drag become inclined as in the diagram, the two combining to neutralize the weight which is supported, as it were, on two inclined strings (Fig. 84). As the nose is pushed farther and farther downwards, the speed of the glide and the angle of glide are both increased until the nose-dive position is reached. When in this condition the whole of the weight is acting as thrust and the whole of the drag is opposing it, there being no longer any lift.

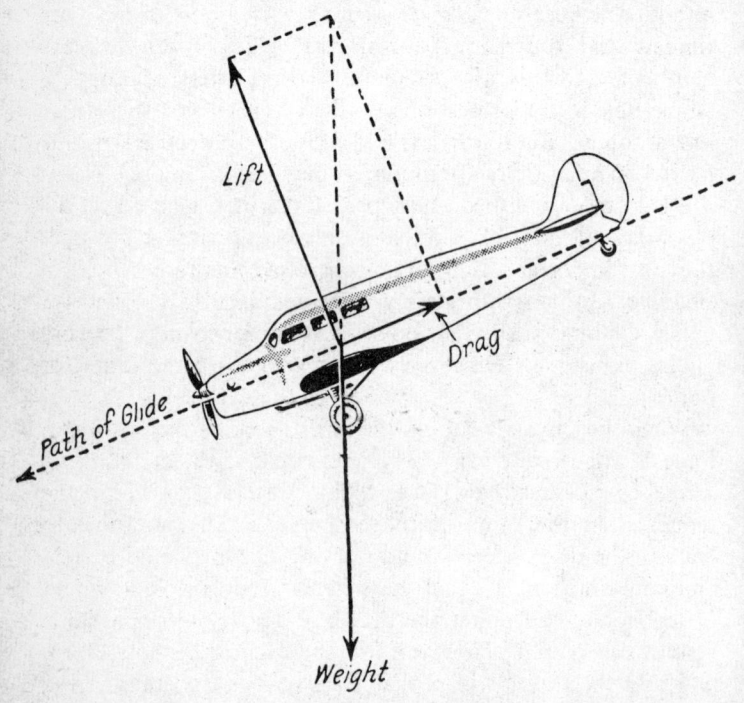

Fig. 84. **Gliding**

But at present we are not concerned with the nose-dive; the main problem in gliding is to travel as far as possible horizontally with the least possible loss of height, in other words to maintain the *flattest possible gliding angle*. As with climbing we must not confuse the gliding *angle* with the gliding *attitude*. The gliding angle refers to the angle of the actual path travelled

GLIDING

by the aeroplane relative to the horizontal. We noticed in the early stages of the book that this angle *appears* to depend on the velocity of the wind, so let us assume, for the moment at any rate, that there is no wind, that the air is perfectly still relative to the earth's surface.

It can easily be shown that the gliding angle depends on the proportions of the lift and the drag. (I have promised you a book without formulae and without mathematics, and sorely tempted as I am, I must keep my promise. But it means that you must take my word for the results instead of understanding them for yourself.) The greater the lift compared with the drag, the greater the lift/drag ratio as it is called, the flatter will be the gliding angle.

Now, we talked a lot about lift and drag, and lift/drag ratio in the earlier stages of this book, but even so I am afraid that some readers may still think of them as rather abstract technical terms. Let us try to realize just what lift/drag ratio means. Lift is what we try to get, drag is what we cannot help getting. The ratio of lift to drag, the ratio of what we want compared with what we do not want, is surely, then, a measure of the efficiency of our aeroplane, so long, at any rate, as we think of the aeroplane as a means of obtaining the maximum amount of lift with the minimum amount of drag. A well-designed *wing* may have a lift/drag ratio, an efficiency, of as much as 20 to 1; it will give 20 lb of lift for every pound of drag. Even higher ratios can be obtained if some method of controlling the boundary layer is employed (see Section 30). Of course, the lift/drag ratio of an *aeroplane* will not be as good as this, because all the other parts of the aircraft—fuselage, undercarriage and so on—add quite a lot to the drag but contribute practically nothing to the lift. Thus the drag may easily be doubled, and the lift/drag ratio, or efficiency of the aeroplane, be reduced to 10 to 1. This is quite a good figure for a well-designed aeroplane.

Thinking of lift/drag ratio as efficiency of design, and going back to the connection, already stated, between gliding angle and lift/drag ratio, we shall see that the more efficient an aeroplane, the flatter is the angle at which it will glide. This is an important conclusion, and it provides us with such a very simple practical means of finding out the merits of a design.

If, for instance, we go to some pains to clean up the design of an aeroplane, to remove unnecessary excrescences, to streamline others, and so on—with a view, perhaps, to entering for some speed contest—we ought to be able to test the effects of our work by simply trying out the aeroplane in a glide—if the gliding angle is improved (i.e. flatter), then the design is improved.

These arguments explain why efficiency of design is of such supreme importance in a glider or sailplane, aircraft which for their success are entirely dependent on having the flattest possible gliding angle. In a sailplane we may sacrifice manoeuvrability, even strength, for the sake of obtaining a good lift/drag ratio. Sailplanes are nearly always monoplanes. They have tapered wings, large aspect ratios, streamlined fuselages of small frontal area, and little or no undercarriage—all these are steps towards efficiency of the kind which affects gliding angle. Also devices which increase the lift, e.g. slots, camber flaps, high-lift aerofoils and so on (provided they do not increase the drag in the same proportion) will all tend towards flatter gliding angles.

What has been said about sailplanes applies equally, perhaps even more, to the man-powered aeroplane (Plate 59), for the problem of sustained man-powered flight has not yet been solved—therein lies much of its fascination—but it will be, and when it is, it will be an object lesson in *efficiency*, efficiency in the machinery which converts man's strength into thrust, efficiency in the external shape and design of the aeroplane,

GLIDING

and efficiency in the internal structure and the materials of which it is built.

We may seem rather to have taken it for granted that a flat gliding angle is necessarily an advantage. For sailplanes it certainly is so, and it will give power-driven aeroplanes a wider range in which to choose a landing-place in case of a forced landing. A pilot's rough rule is that he can glide a mile (horizontally) for every 1,000 ft of height (vertical); this is a poor result, implying a lift/drag ratio of a little over 5 to 1; but it is on the safe side, and he must not forget that the horizontal distance which can be reached will be reduced if he has to glide against the wind.

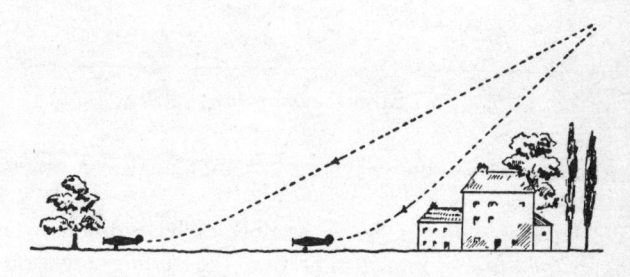

(a) Advantage of the steep gliding angle.

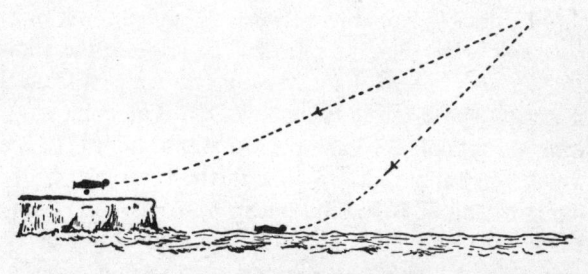

(b) Advantage of the flat gliding angle.

Fig. 85. **Gliding angle**

The flat gliding angle, however, may have certain disadvantages unless, of course, the pilot can choose his gliding angle at will. An occasion on which a steep gliding angle is required is when approaching a small aerodrome or field which has high buildings or trees at its borders. After clearing these obstacles, contact with the ground should be made as soon as possible; otherwise it may not be possible to pull up before reaching the far side of the aerodrome. Fig. 85 shows that on

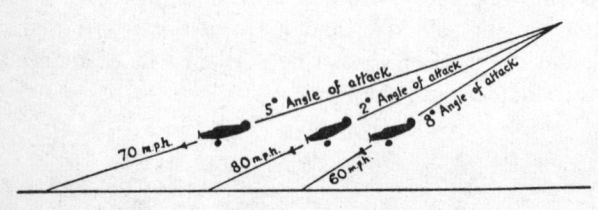

Fig. 86. **Attitudes and paths of glide**

some occasions we may want a flat gliding angle, on others a steep one.

We have seen that *different* aeroplanes have different best gliding angles; but can a pilot vary the gliding angle on a *given* aeroplane? Within the limits, the answer is yes. The ratio of lift to drag will reach its maximum value at one particular angle of attack—probably between 3° and 6°. At angles both *above and below* this, the ratio will be less and the gliding angles steeper. As in level flight, the pilot thinks of it from the point of view of speed rather than angle, but it all comes to the same thing. For him, for example, 70 m.p.h. may be the air speed which will enable him to glide farthest; he *can* glide at both 60 m.p.h. and at 80 m.p.h., but in both cases the path of descent will be steeper. Looking at it from the point of view of angles: 70 m.p.h. may correspond to 5° angle of attack, 60 m.p.h. to 8°, and 80 m.p.h. to 2°. Fig. 86 gives an idea of the attitude of the aeroplane and path of glide in each case.

This may prove very interesting to the pilot who tries it out for himself. Most pilots, if they are trying to reach a distant field, will instinctively try to glide too flat just as they attempt to climb too steeply. If one wants to glide far, it is natural to hold the nose up; but it does not pay. Some pupils overdo it so much that they actually stall on a glide—they reach a 15° angle of attack instead of 4° or 5°.

It is clear that a pilot can intentionally cause the aeroplane to descend more steeply, either by holding the nose up too much or by putting it down too much. Both methods, however, have their disadvantages; the former causes a tendency to stall and to lose lateral control and stability, while the pilot's view of the airfield is very poor when the aeroplane is in this attitude. The latter method—with the nose down—results in a high-speed glide, and although the aircraft comes down steeply it will travel a long distance horizontally to lose its speed just before and just after landing. A better way is probably to alter the lift/drag ratio by artificially increasing the drag by means of air brakes. This device was mentioned when considering landing. We then thought of it as a means of decreasing the length of run after landing, but air brakes may also be used to steepen the gliding angle without endangering lateral control on the one hand, or increasing air speed on the other.

This is an age of flaps, camber flaps, split flaps, slotted flaps, and flaps named after their inventors. For each some special merit is claimed, but all are means of controlling the lift/drag ratio of a wing or of the aeroplane, some improving it by increasing the lift, thus giving lower landing speed, others making it worse by increasing the drag, thus giving a steeper gliding angle without unduly increasing the gliding speed. There are others that claim to do both. The probability is that we have not yet reached finality in the design of flaps and other similar devices. The tendency towards high wing loadings is putting up gliding speeds, better streamlining is still flattening gliding angles, and more accurate judgment is being required from the pilot. He requires better flaps to help him out of his difficulties. The demand is clear enough—no doubt the designers will meet it.

We have so far assumed that there has been no wind, but we must keep in mind that a head wind will reduce the forward speed over the ground compared to the downward speed, thus increasing the steepness of the glide *when seen from the ground* and reducing the distance that the aircraft can glide in that direction. A following wind will have the opposite effects, flattening the glide and increasing the distance. So much for horizontal winds; those of us who spend most of our time on the earth's surface are inclined to forget that the so-called wind is only a part of a circulation current in the atmosphere. When there is a horizontal wind there is also somewhere an *up current*, then more horizontal winds in different directions, and finally a *down current*. The up current is the delight of the sailplane enthusiast; it may even enable him to glide upwards —*relative to the ground*. He may also travel backwards— *relative to the ground. Relative to the air*, sailplaning is merely gliding—*downwards and forwards*, as in all gliding. Up currents may be part of the ordinary circulations of the atmosphere, but they are sometimes very violent under certain types of cloud, or they may even be produced by the wind blowing against the slope of a hill and being deflected upwards.

The air often plays some very curious tricks quite close to the ground. Up currents may be caused by the sun heating a tarmac or concrete road, or a sandy piece of ground, and the aeroplane may feel a "bump" as it comes in to land. But the fact that hot air is rising means that cold air is coming in to take its place and so the pilot may encounter a sudden increase in air speed, or decrease, according as to whether he meets the extra rush of air or finds it travelling with him. In the former case he will feel the extra lift and tend to "balloon" across the aerodrome; in the latter he will lose some air speed and tend to stall. Similarly, the wind blowing over the top of a hangar, or over the brow of a hill, may become turbulent and even blow in the opposite direction. Extreme examples of this kind have been noticed near the Rock of Gibraltar. A further phenomenon is that of *wind gradient*. This is an effect similar to that of the boundary layer over an aerofoil. Even when a considerable wind is blowing twenty or thirty feet above ground level, the air very near to the ground is almost stationary and its movement increases gradually until reaching its full force. Thus the aeroplane, gliding into land, is encountering less and less head wind as it descends through the last few feet—it is losing air speed, and may tend to stall. The moral of all this— from the pilot's point of view—is to glide in more steeply and at higher speed than he would otherwise do, to keep some speed in hand for any emergency. One must remember that the pilot *cannot see* these up currents, bumps, wind gradients or whatever they may be.

Throughout the discussion on gliding we have assumed that there is no thrust from the engine. In the old days it was considered bad flying to use one's engine during the approach, presumably because it made things too easy or, perhaps more justifiably, because one never knew when the engine would stop and one would have to land without it. Today we are not so silly as to be ashamed of making things easy—if we know how to do so! Thus it is that the *engine-assisted approach* has become a recognized method of coming in to land. There are several advantages.

The aeroplane is more controllable and less margin of speed is required to overcome such things as bumps and wind gradients. The attitude is such that very little change is required for the actual process of landing or—equally important—for making another circuit in case the distance has been badly judged. The engines, already running, respond more readily to sudden changes of the throttle, and there is less change of trim when they do so respond. The reader may well ask, but will not the engine-assisted approach mean an even flatter gliding angle, and have we not already said that a flat gliding angle is a disadvantage? That is true—we never seem to get everything we want in this subject.

Earlier mention of sailplanes and up currents leads us to an important aspect of gliding that we have not yet considered, and which cannot be omitted. We have said that the flattest possible gliding angle depends on the lift/drag ratio, and people are apt to say, "But what about weight?" If one tells the truth and says that the weight of the aircraft doesn't affect the issue, such people are apt to become quite indignant and to say that it obviously—dangerous word—does. One has even heard the retort "What about a brick?" To this one is tempted to reply, "Well, what about a brick?" No, a brick hasn't got a good gliding angle, but this is not because it is heavy (incidentally it is lighter than any aeroplane), but because it just hasn't got any lift/drag ratio; fit it with a nice pair of wings, streamline it a bit, and it will glide beautifully. But why this prevalent idea that weight must affect the best gliding angle? It is a simple fact that some of the heaviest of aeroplanes, even those with the highest wing loadings, can glide the farthest from a given height. "Are you going to tell me then," someone says, "that it doesn't matter how heavy a sailplane is? Why, surely . . ." Yes, yes, don't get excited, *of course* it matters how heavy a sailplane is; it matters how heavy any aeroplane is—haven't we said so all along? The weight affects all kinds of things, how slowly it can fly, and as we discovered in Section 61, how far it can fly, how high it can fly, how quickly it can climb; in fact, practically everything except the flatness of its gliding angle. The weight, or rather the wing loading, has a most important effect on gliding. "Ah," they interject with some relief. Yes, *the less the wing loading* the more slowly can the

GLIDING

aeroplane glide and the *lower is its rate of descent*. That's the clue. Sailplanes are made as light as possible, not to give them a flatter gliding angle (that can only be done by streamlining), but to enable them to lose height less rapidly, or, what comes to the same thing, to gain height even in slight up currents. Other aeroplanes are made light for even more important reasons.

At this stage one must fall into the temptation of putting to the reader a problem of flight which, though it doesn't often

Fig. 87. **Can you make it?**

arise in all its details in practical flying, is such a good test of one's understanding of the principles of flight—and particularly of gliding—that it should be considered by everyone who is trying to understand those principles. Here, then, is the problem—*suppose one is flying an aeroplane well out to sea when the engine fails; there is a good airfield just on the coast and it is touch and go whether one can reach it; you have disposable load on board, luggage, bombs, fuel, passengers!—or whatever it may be; should you jettison your load, and if so when, and what should be your tactics in an endeavour to reach the airfield?* (Fig. 87). Now, it is not an exaggeration to say that a book could be written in answer to that question—in fact, quite a lot of this book is concerned with the answer in one way or another. For this reason, I am going to play rather a dirty trick and, having posed the problem, leave the reader to answer it—but not without giving him a few hints and help by suggesting some of the things that must be considered:

(a) The direction and strength of the wind makes all the difference—take first a condition of no wind.

(b) When your engine fails you will probably be flying at a speed higher than that of your speed for flattest glide—what can you do with this surplus speed?

(c) What about drag? Can you get rid of any? What are possible suggestions?

(d) What about flaps? Should you use them or not? Yes, it all depends—but depends on what?

(e) At what speed should you fly?

(f) Should you jettison any load? If so, when? This is the most important part of the question, and the answer is not quite what you might think from what has been said earlier in this section. Remember that load affects speed of glide, but not angle of glide—remember, too, that it affects landing speed! —Yes, both these facts affect the answer.

(g) Now imagine that you have the help of a following wind —and go through all the consideration again—does it affect b?, c?, d?, e?, f?. Just to prevent you dismissing these too easily, I will give you the hint that in most of them the answer is "Yes." And a further and most important hint is that with a following wind all the time that you stay in the air the wind is helping you, so . . . And another hint, and one that hasn't really been mentioned previously—at a certain speed you will get the flattest glide (in no wind conditions), *but at a rather lower speed (wind or no wind) the aeroplane will have rather less rate of sink; in other words it will stay in the air longer. So which, if either, speed is it to be with a following wind?*

(h) Now a head wind. Go through them all again. If you consider them aright, you will find that your tactics should be quite different.

(k) Now add wind gradients to g and h and the effect of the cliffs on the wind. In practice, these may make all the difference not only to your tactics, but to the chances of reaching dry land.

You could go on to consider side winds, but you have probably had enough, and if you have really tried to think of all the factors involved, you will have learnt a lot about the problems of gliding—and, if you ever find yourself in such a predicament, you may just remember one or two points that will help; *you certainly won't have time to work it all out in the air.*

69. Climbing

An aeroplane can climb if it has power in hand over and above that required for straight and level flight at the particular height at which it is flying. More power is required as greater heights are reached. This last statement may sound rather surprising, because it was stated in Section 23 that the drag depended on the air density, which decreases with height. Unfortunately, however, at whatever height we fly, the lift must be kept equal to the weight. Now, the lift also depends on the air density, so that we must make up by other means the lift that we lose due to the decreasing air density. "Other means," as in Section 65, may consist of an increase in:

(a) *Camber of wings*   (c) *Wing area*
(b) *Angle of attack*   (d) *Velocity*

Each of these means an increase in drag, which in turn means greater power required. But it is difficult to attain extra power at high altitudes; in fact with propeller propulsion and the ordinary piston engine the tendency is definitely in the other direction, the power falling off as greater heights are reached. With this falling power and increasing drag it is almost inevitable that velocity must be sacrificed and the lift made up by increasing (*a*), (*b*) or (*c*), of which only (*b*) is possible on most aeroplanes.

It should be clear from these arguments that as height increases more power is required for level flight, probably less power is available, and therefore the *extra* power needed for climbing falls off rapidly. Thus the *rate of climb* decreases as we get higher, until eventually all the power available is used for level flight and further climbing is impossible. The aeroplane has then reached its *absolute ceiling*. In practice it takes much time and patience to reach the *absolute* ceiling, because the nearer we get to it the more difficult does it become to get any farther. For this reason the *service* ceiling is used for tests and specifications of aeroplanes. This is the greatest height to which the aeroplane is ever likely to be climbed in ordinary use, i.e. excluding such things as attempts to beat height records, and it is usually defined as the height at which the rate of climb has become reduced to 100 ft/min. This contrasts with the 3,000, 4,000 or 5,000 and more feet per minute of modern aeroplanes, near ground level.

It is not at all easy for a pilot to judge at what attitude he should maintain his aeroplane in order to achieve the greatest possible rate of climb. Suppose he is flying level at 100 m.p.h.; he has power in hand and opens his throttle until the engine is developing maximum power. If he leaves the aircraft to itself, it will start to climb; but if he puts it into a new attitude, with the nose more downwards, he can keep in level flight at a higher speed than normally; in fact he can thus achieve his *maximum speed of level flight*, for the sake of argument let us say 150 m.p.h. If he then pulls his control column back slightly, the elevators will be raised, the nose of the aircraft will come up slightly, and it will begin to climb at a forward speed of less than 150 m.p.h., say 130 m.p.h. The rate of climb may

CLIMBING 165

be, say, 200 ft/min (Fig. 88). When climbing like this at high forward speed, the nose of the aeroplane is not pointed upwards at all, and the angle of attack will be even less than when in level flight at 100 m.p.h.

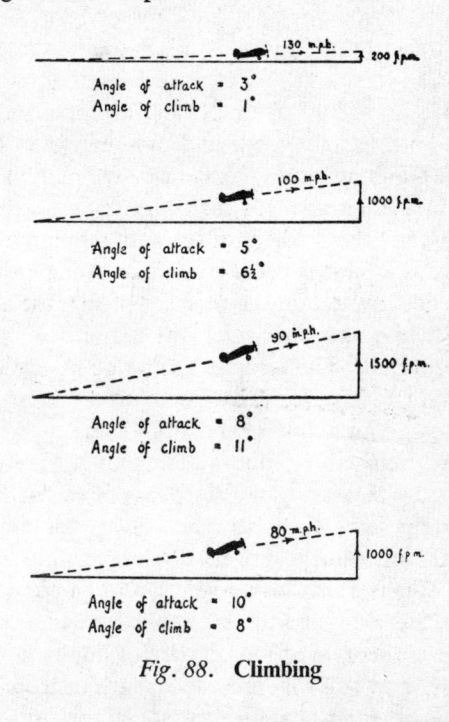

```
                    130 m.p.h.         200 f.p.m.
Angle of attack = 3°
Angle of climb = 1°

                    100 m.p.h.         1000 f.p.m.
Angle of attack = 5°
Angle of climb = 6½°

                    90 m.p.h.          1500 f.p.m.
Angle of attack = 8°
Angle of climb = 11°

                    80 m.p.h.          1000 f.p.m.
Angle of attack = 10°
Angle of climb = 8°
```

Fig. 88. **Climbing**

Now suppose the pilot continues his experiment and pulls and pulls the nose up a little farther. Forward speed will drop to say 100 m.p.h., rate of vertical climb increase to say 1,000 ft/min, the aeroplane will be slightly more in a climbing attitude, and the angle of attack will be increased but still remain quite small, probably about the same as it was in level flight at the same speed.

Now again: speed say 90 m.p.h., rate of climb 1,500 ft/min, attitude steeper, path of climb steeper, angle of attack of wings on the air greater.

This process does not go on for ever. It is quite possible, for instance, that at 90 m.p.h. we have reached the greatest rate of climb, namely 1,500 ft/min. If this is so and the experiment is continued, the result will be a forward speed of say 80 m.p.h., and rate of climb 1,000 ft/min. The attitude and angle of attack will both be increased, and although the aircraft will *appear* to be climbing more steeply (even the pilot may imagine this if he does not watch his instruments), actually both the rate of climb and the gradient of the path of climb will be less. If the reader is a pilot, let him try this out for himself; if one can in practice investigate the theories of flight it adds a new interest to flying, and one which the majority of pilots have not yet discovered. The aircraft should be fitted with an *air-speed indicator* to give forward speed, an *inclinometer* to give the attitude, and an *altimeter* and stop-watch or, better still, a *rate-of-climb indicator*. But unfortunately there are no instruments for measuring the steepness of the path of climb, which is *not* the same as the attitude—a very common mistake. Nor does any instrument give us a direct reading of the angle of attack. This is a pity, as we said before in connection with stalling, because no instrument could be more valuable in linking up the theory and the practice of flight.

The reader may ask why there is no such instrument. There are two reasons. First, that no accurate and reliable instrument for the purpose has been devised, and secondly, that such an instrument is not essential as a guide to practical flight, because of the close connection between angle of attack and air speed. In level flight, for instance, there is one, and only one, angle of attack for each particular air speed for a given aeroplane. Although the same relationship between angle of attack and air speed may not apply exactly to climbing and gliding, the

CLIMBING

fact remains that a scale of angle of attack could be marked on the dial of an air-speed indicator and could be used as being reasonably accurate. But if this is so, the air speed might as well be used instead of the corresponding angle, and thus it is that the pilot thinks and talks of air *speed*, of stalling *speed*, best climbing *speed*, best gliding *speed*, and so on, whereas those more concerned with the theoretical side of the question think and talk more of *angle* of attack, stalling *angle*, and *angle* of attack for best climb and best glide.

But to return to our climbing experiment. We had reached 80 m.p.h. and a rate of climb of 1,000 ft/min. Further pulling up of the nose will result in a steadily decreasing air speed, a decreasing rate and angle of climb, and an increasing angle of attack. In time we shall reach a point when there is no longer any climb, and then we shall be in level flight again, but a very different level flight from that at which we started. The nose is pointing up in the air, the engine labouring, with the angle of attack near the stalling angle and the speed the lowest speed of flight, or stalling speed, say 45 m.p.h. Notice, once again, that this minimum speed of flight—like the maximum speed—is attained with the engine running at full throttle.

Let us pause for a moment at this idea of flying at the stalling speed, the aeroplane being inclined at a large angle and the poor old engine labouring away in its attempt to drag the aircraft along and to keep it airborne. A rather interesting question arises. Is it, by any chance, possible to fly beyond the stalling angle or below the stalling speed? One would think that the answer must be No—have we not emphasized the fact that this is the lowest speed of flight, the largest possible angle of attack? But we get many surprises in studying this subject—and a very clever pilot may succeed in flying beyond the stalling angle, even up to angles of attack such as 45°. After all, an aerofoil does not by any means lose *all* its at lift the stalling angle, and at 45° the lift is still quite

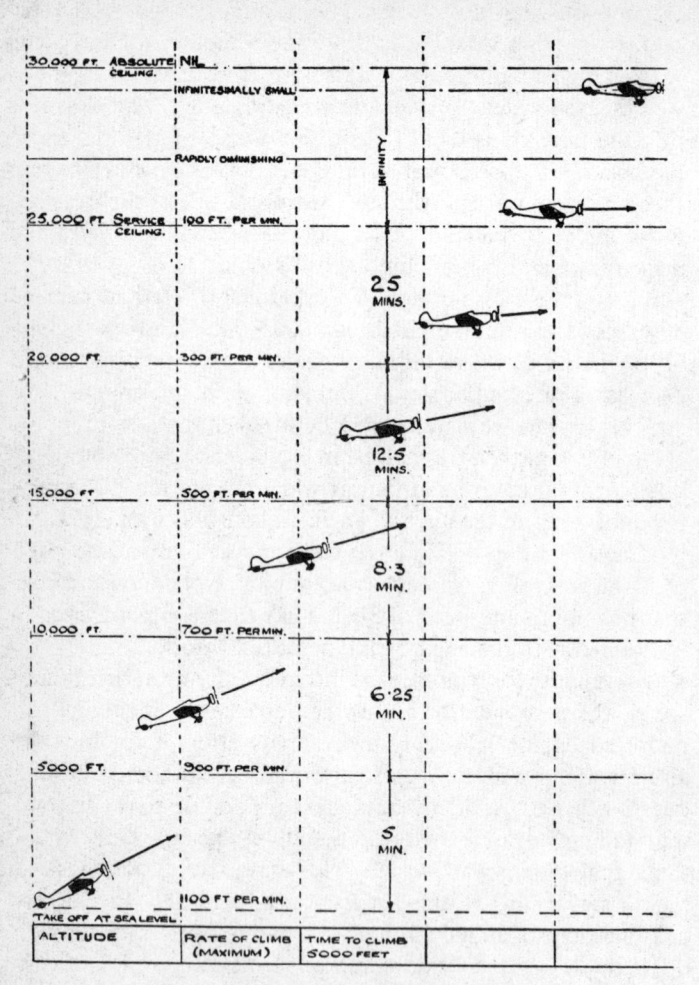

Fig. 89. History of a climb to the ceiling

CLIMBING

appreciable (although the drag is very great), and the large inclination of the aircraft means that the thrust of the airscrew is itself providing quite a large share of the lift. The major difficulty is the inadequacy and inefficiency of the control at these angles and speeds—hence the necessity for the "very clever pilot". Also very great engine power is needed, since the drag is so great, and this large expenditure of engine power is accompanied by inefficient conditions for cooling the engine owing to the low speed of air flow. Altogether, it is hardly surprising that the average pilot thinks of stalling as the limit of flight. Flight in the stalled state can hardly be considered as a satisfactory means of locomotion—perhaps that is why Lewis Carroll attributed it to the strange animal the Jabberwock which "burbled as it came"!

Now let us sum up our experiment. For this particular aeroplane, climb at full throttle is possible at all air speeds between 150 m.p.h. and 45 m.p.h., but the greatest rate of climb is at 90 m.p.h. As might perhaps be expected, there is a tendency on the part of pilots to hold up the aeroplane too steeply when climbing, especially if they do not know the best climbing speed. You will remember that, in just the same way, they tended to glide at too flat an angle. Aeroplanes do not always travel in the direction in which they are pointing.

So far we have assumed that our typical aeroplane is climbing near the ground; how will the position change at greater heights? As already explained, the power in hand will become less, the rate of climb will fall off at whatever air speed one attempts to climb, and the *range* of available speeds both for level and climbing flight will also contract. Maximum speed will decrease, minimum speed will increase, until eventually the two will become one and the same speed, the only possible speed for level flight at this height, when no further climb is possible. The absolute ceiling has then been reached. Fig. 89 illustrates the history of a climb to the ceiling. It is not easy in such a diagram to indicate clearly the distinction between the attitude of the aeroplane *relative to the ground* and its attitude *relative to the path of climb* (or angle of attack). Although the former decreases with height the latter will *increase*, being greatest at the ceiling.

No mention has so far been made of a *supercharger* on the climb. The supercharger is a device fitted to an engine which blows extra air into the cylinders at altitude to compensate for the decrease in the density of the air. In this way the power delivered by the engine may be maintained up to a considerable height, thus increasing the rate of climb while the supercharger is in use. By such means the rate of climb, ceiling and service ceiling may all be increased. The turbine engine normally used for jet propulsion is in itself a type of supercharger, and for this and other reasons jet propulsion shows up much better than propeller propulsion when it comes to performance at height.

The climbs which we have considered have been made at full throttle, a state of affairs which one does not care to maintain for long, and therefore it is interesting to examine the effects of throttling down to lower engine speed. The results are similar to those caused by altitude. As we throttle down, the minimum speed may increase slightly, the maximum speed of level flight will decrease considerably, and the loss of surplus power available for climb will result in a diminishing rate of climb, until eventually there is no climb. When this point is reached there is only one speed for level flight, and the engine is running most economically. Conditions are ideal for ordinary cruising; we are in fact at the speed for best endurance as described in Section 61. In practice, the term *cruising speed* is rather vague—for, as we have seen, one may cruise at a speed to give greatest economy of fuel, or to give maximum range of action, or to give greatest comfort in flight. All are, in a sense, cruising speeds—yet all are different.

70. Turning

When anything travels on a curved path there must be a force acting on it in a direction towards the centre of the curve; otherwise it would continue to travel *in a straight line* in the direction in which it happened to be going at any particular moment. The force towards the centre—called *centripetal* (centre-seeking) force—is provided differently in the various means of transport. In a ship it is the pressure of the water on the side of the hull, namely the side farthest away from the centre of the curve. The ship, in its turn, presses back on the water—this is the equal and opposite reaction, the outward force, called *centrifugal* force. It is rather curious that this term is much more widely known than centripetal force, for it is the centripetal force which acts on the body, which is what really matters. When a motor car turns a corner on a road that is not banked, the inward force is provided entirely by friction between the tyres and the road surface, and all motorists know (sometimes to their cost) that if this force of friction is not sufficient, no amount of turning the steering wheel will coax the car round the corner. But neither the ship nor the car on the unbanked road turns a corner really satisfactorily; in both instances outward "skidding" tends to take place. The best way for a machine, be it aeroplane, train or car, to turn a corner is by *banking*. The ship is rather a problem in itself, usually tending to bank the wrong way, so let us leave it out of the discussion.

The idea of banking is to eliminate the tendency to skid. In other words, we must not rely on friction to provide the inward force. When the aeroplane, train or car is travelling on a straight course, it receives upward forces from the air, rails or road respectively. Therefore when turning a corner there must be both an *upward* and an *inward force*, in short a resultant and larger force acting upwards and inwards. By banking to the correct degree, the aeroplane presents its wings at right-angles to this force so that there is no tendency to skid outwards as there would be with no banking, or to slip inwards as there would be with too much banking.

In Fig. 90 we have put in the centrifugal force because this is quite a good way of thinking of it, provided you are quite sure

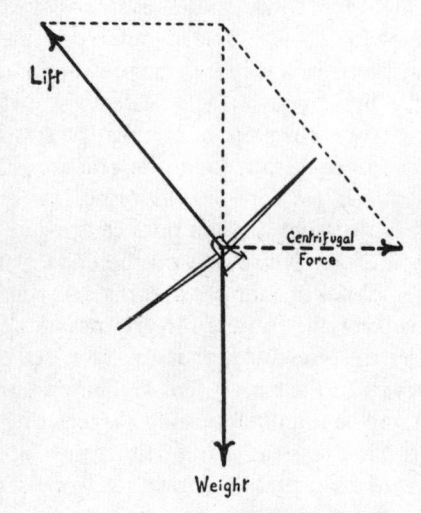

Fig. 90. **Turning**

that you understand what it really means. A body travelling on a curved path is all the time *accelerating* towards the centre, and so the forces acting on it are *not balanced*; but if we put in a centrifugal force *and* think of it as being a simple problem of statics or non-accelerated motion, we are still correct. Some people say that all proverbs are untrue, and here we seem to have an instance where the safest of all proverbs breaks down, because apparently two wrongs *do* make a right.

The train and the car are not so fortunate as the aeroplane because they cannot choose their bank; the rail and the road

TURNING

are already laid for them. Now, the amount of bank required obviously depends on the relative values of the upward force and the inward force needed. The upward force must be equal to the weight, as in normal flight; but the inward force increases with speed and decreases when the radius of the curve is large. Thus the faster the aircraft, and the smaller the radius on which it turns, the greater will be the force required towards the centre and the steeper should be the bank. Clearly, therefore, the banked railway track or road corner can only be suitable for one particular speed, since its radius and angle of bank are fixed. The only parallel to the freedom of action possessed by the aeroplane is the car on a racing track, where there are usually different degrees of bank, which the driver can choose according to the speed of his car. Even so, the radius of turn is fixed, whereas the pilot of an aeroplane can choose at will the radius, the speed and the angle of bank.

For these reasons there is really no excuse for turning with an incorrect bank, although it is perfectly possible to do so. If the bank is not correct, the aircraft will either skid outwards or slip inwards; the former indicates too little bank, the latter too much bank. The pilot has several means of knowing whether the angle of bank is correct (Fig. 91). In an open cockpit, if there is a skid outwards, wind will strike his face on the outside of the turn. If the aircraft slips inwards, wind will come from the other side. A piece of string or streamer of any kind exposed to the air flow will indicate incorrect bank, since it will be affected by the wind caused by side-slipping. For instance, if the streamer blows outwards, there is too much bank. A pendulum or plumb-bob hung in the cockpit—not exposed to the air—is another indicator of side-slip, i.e. of incorrect bank. On a correctly banked turn it will hang centrally *relative to the cockpit*, as when flying straight; if over-banked, it will be inclined inwards; if under-banked, outwards. The slip needle on the so-called *turn-and-bank*

174   FLIGHT WITHOUT FORMULAE

indicator (an instrument often used in modern aircraft) is connected to a pendulum, and is really a side-slip indicator—notice that it is made to move in the same direction as the pendulum. The turn needle of this instrument shows the rate

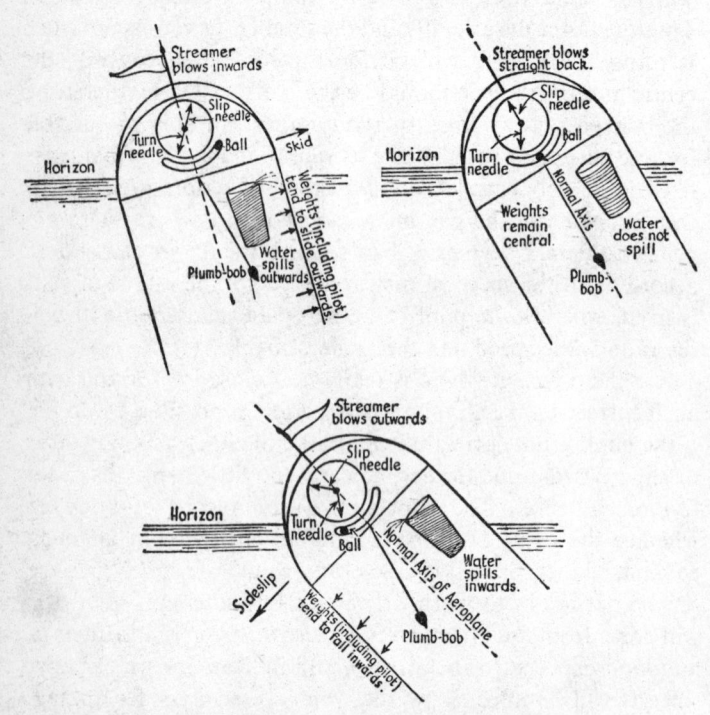

Fig. 91. **Effects of correct and incorrect bank**

of turn, and this, of course, will be in the same direction whether the bank is correct or not. But the simplest and most reliable instrument is the ball in a kind of inverted spirit level. In the good old days an actual spirit level was used and the pilot learnt to "follow the bubble" with his stick, and so keep the

TURNING

bank correct. But the bubble was just air, and it was the liquid which had the weight, whereas in the modern instrument the ball is the weight and so moves in the opposite direction to a bubble—it moves, of course, in the same direction as all other weights, including that of the pilot, and the stick must be moved in the *opposite* direction in order to correct the bank; this is a more instinctive action than with the old bubble. With all these guides and the pilot's own instinct, which ought to detect any side-slip, there does not seem any excuse for incorrect banking.

Since the wings must provide both the upward and the inward forces, they are more heavily loaded during a turn than in straight flight. At a banking angle of 60° the loads will be twice normal, and at 84° ten times normal. We are not seriously concerned in this book with the effect of these increased loads on the structure of the aeroplane, but perhaps we ought to consider the pilot who will also experience them. At 84° he will sit on his seat with ten times his normal weight; but what is perhaps more important is that all the loose things inside him, especially his blood, will tend to move from his head towards his feet. The effect varies with different pilots, but with all it results sooner or later in a temporary "blacking-out," or blindness, and almost total lack of consciousness. It passes off as soon as the bank is completed, but it sets a definite limit to the suddenness of turns at very high speed, and it is obviously no use making the structure very much stronger than the pilot.

It is perhaps unfortunate that pilots have learnt to talk of the extra loads in turns and other manoeuvres in terms of g. Unfortunate for two reasons. First, because it needs a certain amount of explanation; secondly, because it must be admitted that even after such explanation all too many pilots are very vague as to its meaning, and often reveal as much in their speech. In order to produce a change of motion there must be a force. A change of motion (or *acceleration*, as it is called in mechanics) means any change in speed or any change in direction, the latter being the more important in flying. Now the greater the acceleration, the more sudden the change of motion, the greater must be the force producing it. If we drop a body, it is acted upon by a force equal to its own weight, and we know by experiment that if there is no air resistance it will increase its velocity, i.e. accelerate, at the rate of 32 ft per second every second. For simplicity we call this acceleration g—the acceleration of gravity. Now, when aeroplanes manoeuvre, as for instance in a turn or in a loop, or pulling out of a dive, the accelerations *in the line of flight* are not very large, that is to say, the speed does not increase or decrease very rapidly. But *at right-angles to the line of flight*, i.e. in the same direction as lift, the accelerations may be very large, two, three or more times the acceleration of gravity, or, in short, $2g$, $3g$ or more. These high accelerations are due to changes of *direction* at high speed. And, as already stated, such accelerations need proportional forces to produce them, and so it is that the loads in manoeuvres go up to two, three or more times the normal load.

To the forces necessary to produce the acceleration must often be added the normal weight of the aircraft, as, for instance, at the bottom of a loop, or when flattening out after a nose-dive. If, in such circumstances, there is an acceleration of $2g$, the loads will be three times normal, twice to produce the acceleration *plus* the normal load to lift the weight. The pilot—incorrectly—often speaks of this as $3g$. The *acceleration* is only $2g$, it is the *load* which is *three* times normal.

It is interesting to note that the average pilot will "black-out" at accelerations of about $4g$ to $5g$, that he may suffer actual physical injury at $7g$ or $8g$, and that the average aeroplane will show signs of breaking at $9g$ or $10g$.

Another problem of the high loads involved in turns and other manoeuvres is that extra speed is necessary to produce these loads: in other words, the *stalling speed goes up*. For a

TURNING

load four times normal (e.g. at an angle of bank of 75½°) the stalling speed will be double that of level flight; for a load of nine times normal (e.g. at an angle of bank of 83°) the stalling speed will be three times that of level flight. The pilot becomes so accustomed to thinking of the stalling speed of an aeroplane as being say 60 m.p.h., that he is apt to forget that it may stall at 120 m.p.h., or even 180 m.p.h.—in a steep turn, or when pulling out of a dive.

After considering these points and noting the tremendous increase in load between 75½° and 83°, the reader may well ask what will be the loads, and what will be the stalling speed, in a 90°, or vertical, bank. The answer is that, according to the ideas we have been working on up to the present, a vertical bank is not possible—*without losing height*. When the wings are vertical, the lift will be horizontal, and no horizontal force, however large, can lift the weight of the aeroplane. In case the reader should himself be a pilot—and I hope he is, because it is for such as him that this book is written—let me hasten to explain that I am not saying that it is impossible to bank an aeroplane vertically. The pilot knows perfectly well that it is possible; the point is simply that it is a new condition of flight, and one which we have not yet considered. It is *not* just an ordinary banked turn, only more so. In practice, truly vertical banks are not very common; a steep bank is one of those things which seem to be much more than they really are, and it is easy to believe that the bank is vertical when it is, in fact, no more than 70°. Thus we must realize that we hear about far more vertical banks than really happen. True vertical banks may occur in the following conditions.

Momentarily, at a certain part of a steep turn, the pilot may bank vertically, or even over the vertical, on the same principle as looping the loop, a loop being, in this sense, a bank of 180°. Under such circumstances there need be no side-slip, *but the bank is not constant round the turn.*

A continuous vertically banked turn may be executed, the aeroplane side-slipping downwards all the time. The side-slip, owing to the directional stability, will make it difficult for the pilot to hold the nose up. There will, of course, be a considerable loss of height during the turn.

The third possibility, and probably a very rare occurrence in practice, is a vertical bank without loss of height or side-slipping. The secret of this is that the nose is inclined upwards, and sufficient lift is obtained from the upward inclination of the thrust, and from the air flowing over the inclined fuselage, to balance the weight. It sounds rather doubtful—and it probably is.

Let the pilot who is sufficiently interested investigate this problem—*at a good height*. The author is anxious to help to fill the gap between the two parties who are keen on aviation—the practical pilots who think that all theory is rubbish and only actual flying really matters, and the large body of theorists and talkers whose knowledge of flight has all been gained on the ground. Strange as it may seem, there is something to be said for both parties, and the much too wide gap between them would be narrowed down if the pilot were a little less afraid of thinking things out, and the theorist a little less afraid of going into the air.

Until recent years much stress used to be laid on the apparent change in the function of the elevator and rudder controls as the bank is increased. Take the extreme example of a vertical bank. When in such a position it would seem that the rudder would raise or lower the nose of the aeroplane, while the elevators would control its direction of travel. In short, the rudder would become elevator, and the elevator would become rudder. At normal banking angles the change would only be partial. It all sounds as though it would be rather puzzling to the pupil pilot, and indeed it was. Nor were the assumptions involved quite true. In a vertical bank the rudder will certainly *try* to act like an elevator, but one has only to try it to discover that, as such, it will be *singularly ineffective*. There is nothing surprising about this if one realises how an elevator really works; when in normal flight the elevator is raised a downward force is created on elevator and tail-plane, but this in itself will not raise the nose—the reason why the nose goes up is that the angle of attack, and so the lift, on the main planes is increased. Now go back to the vertical bank and try "raising" the rudder; this will not raise the nose except in so far as it can incline the body of the aeroplane and get lift from it. In most types of aeroplane it just will not be able to do this; in fact the chances are that the aeroplane will go into a steep side-slip and the nose will drop in spite of all the efforts of the rudder. In modern methods of training the pilot learns to think of *the elevators as controlling the position of the nose even in quite steep banks*—not, of course, vertical banks. He thinks of *the ailerons*, rather than the rudder, *as controlling the rate of turn*—the ailerons cause the bank, the bank (unless rudder is also used) results in side-slip towards the lower wing, the side-slip (owing to directional stability) results in a turn. What, then, is the function of the rudder? In this method of training the rudder is thought of as a controller of *slip or skid*, always being used in the direction of such slip or skid—and we have already considered how the pilot knows whether he is slipping or skidding. At first people were inclined to think that these modern ideas were revolutionary, but they are nothing of the sort; a good pilot trained on them uses his controls in exactly the same way as a good pilot trained on other methods; the difference is only in the way he thinks about it, and the advantage of the modern thinking is that it involves no change in the functions of the control surfaces as bank is increased.

In case the reader should feel that we have devoted too much space to the ordinary turn, perhaps it should be mentioned that the turn is not only the most common and the most useful manoeuvre, whether in peace or war, but that it is typical of all other manoeuvres. If we understand the problems of the turn, why the loads are increased, why the stalling speed goes up, we are well on the way towards understanding many other things.

71. Nose-Diving

A nose-dive is another of those manoeuvres which feel much worse than they really are. True vertical dives are very rarely performed in practice, but any dive at an angle greater than 70° to the horizontal seems near enough vertical to a pilot—and even more so to his passenger.

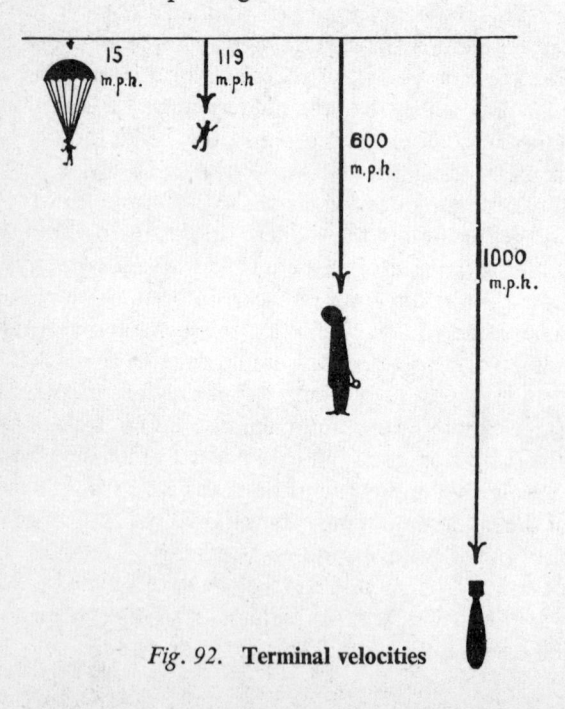

Fig. 92. **Terminal velocities**

NOSE-DIVING

If an aeroplane is dived vertically, it picks up speed until it is in a state of equilibrium. This will occur when the drag (now the upward force) equals the weight (the downward force). When this condition is reached, the aeroplane is travelling at its *terminal velocity*. For modern types of aircraft this will probably be over 400 m.p.h., and may be as high as 600 m.p.h. or more. The speed attained in such a dive is hardly affected by the power of the engine, since it makes very little difference whether the engine is running or not; it depends, of course, on the weight of the aeroplane, since the greater the weight the greater will be the drag required to balance it; but it depends more on streamlining and all the attempts which have been made to reduce drag, such as small frontal area, smooth surfaces, and so on. It is interesting to compare the terminal velocities of the following (Fig. 92):

Man descending by parachute (15 m.p.h.).	Large area of parachute designed to create drag.
Man falling in air without parachute (120 m.p.h.).	Speed kept comparatively low by poor streamlining.
Aeroplane diving vertically (400–600 m.p.h.).	Well streamlined.
Falling bomb (800–1000 m.p.h.).	Heavy, small frontal area, streamlined.

To the aeroplane designer it is from the *structural* point of view that the nose-dive is most interesting. The angle of attack during a dive is very small, or even slightly negative. For this reason the distribution of pressure over the wings is quite different from the conditions of normal flight. The most noticeable difference is that the pressure at the front of the top surface of the wing is increased instead of decreased, and thus we get a downward force at the front of the wing and an upward one at the rear, the two combined tending to turn the aeroplane over on to its back. This twist on the main planes provides one of the problems, but there is another in that the aeroplane is prevented from turning over by a considerable down load on the tail plane, and this puts severe and unusual loads on the tail unit and fuselage (Fig. 93).

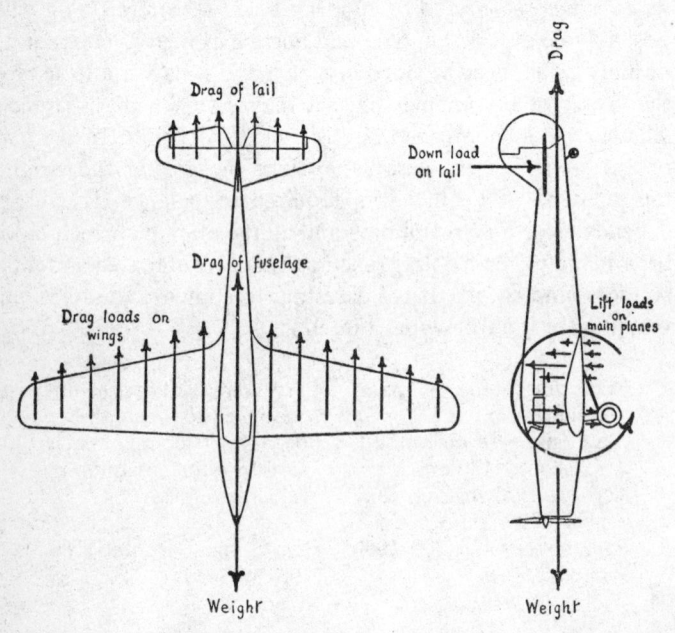

Fig. 93. **Loads during a nose-dive**

A nose-dive is one thing—getting out of it is another. The loads during the dive itself are unusual—and interesting for that reason. The loads experienced when the pilot pulls the machine out of a dive are interesting for quite a different reason. They are similar in nature to the loads of ordinary flight, but *they are very large*. How large will depend on the velocity of the dive and the suddenness of the pull-out. Suffice it to say that they may easily prove to be the greatest loads ever encountered in flight, accelerations may go up to 6*g* or more, the pilot must beware of blacking-out and even of breaking his aeroplane, and stalling speeds may be reckoned in hundreds of miles per hour.

72. Taxying

Movement over the ground is called *taxying*. From the piloting point of view this is not so easy as it may look, but it does not raise many theoretical problems. There must, of course, be control over the machine so that it can be manoeuvred into the best position for taking off. The engine and propeller or jets are used to provide the thrust which pulls or pushes the aeroplane forward. Owing to the low speed of movement on the ground, the control surfaces are comparatively ineffective, especially when taxying downwind. A following wind on the ground may give rise to conditions quite unlike those experienced in flight. The rudder and elevators become more effective if the engine is opened up and the high speed of the slipstream is allowed to flow over them. This does not apply to jets, where, for obvious reasons, the jet does not strike the other parts.

Another difficulty of taxying in the tail-wheel type of aircraft (Plates 11, 13, 14, etc.) is that, the centre of gravity being behind the wheels, the aircraft is directionally unstable when on the ground, the slightest yaw being aggravated by the weight tending to swing farther round. This is not so with the tricycle or nose-wheel undercarriage, which is now the standard type on most large aircraft (e.g. Plate 53).

A steerable nose or tail wheel is a great help. Taxying in a strong tail or side wind is difficult, and in small aircraft it may be necessary for mechanics to hold the wing tips in order to guide the aircraft. Two or more engines are a great aid to taxying and may enable an aeroplane to be turned about in its own length. Reversible propellers or jets are even more effective, and wheel brakes, acting differentially on the two wheels, are a tremendous advantage, and are always used on modern aircraft.

73. Taking off

After taxying into the best position, and turning into the wind or along the runway, the next problem is the take-off. In this again, the pilot has a number of things to think of, and it requires skill and practice, but the principles are simple. The aeroplane being head to wind, or as nearly so as the runway in use allows, already has *some* air speed, and as it runs along the ground this air speed increases until the aeroplane becomes airborne. In order to take off in a small space, speed must be gained as rapidly as possible. In a tail-wheel type, the pilot endeavours to raise the tail as soon as he can, thus presenting the main planes at a small angle to the air and reducing the drag of the whole aircraft to a minimum; in a nose-wheel type this, of course, is not necessary, the wings already being at a small angle.

Once the minimum flying speed has been attained, the aeroplane will actually leave the ground sooner if the tail is lowered so that the wings present an angle of about 15° to the air flow. This method might be used in an emergency (such, for instance, as having to clear a ditch), but it is apt to be dangerous in practice because the aeroplane leaves the ground in or near a stalled attitude, the controls being ineffective, the nose cannot be inclined further upwards or the machine will stall, and the only way that speed can be gained is to raise the tail, when there will be a danger of touching the ground again. So it is better to pick up a reasonable speed before leaving the ground, even if one is approaching obstacles which have to be cleared or avoided. With sufficient flying speed one can climb or turn; without it, one can do nothing.

TAKING OFF

By taking off against the wind not only does one start with some relative air speed, and the length of run is thereby shortened, but, perhaps most important of all, the angle of climb (*as viewed from the ground*) is increased and obstacles on the edge of the aerodrome are more easily cleared. In these days of runways it may not be possible to take off exactly into wind; indeed when only one runway is used the wind may even be at right-angles to the take-off run. This means not only that the run will be longer but that the aeroplane, owing to its directional stability, will tend to turn off the runway into the wind, and owing to its lateral stability, will tend to bank away from the wind. It sounds rather alarming, but nowadays pilots learn, even in their early training, the necessary tricks and technique of dealing with the situation, and in practice, cross-wind take-offs—and landings—present little difficulty.

The high wing loading of modern high-speed aircraft is creating new problems in connection with taking off. Such a high speed is needed that rockets are sometimes used to boost the thrust, or external assistance may be applied. Experiments with catapults have been proceeding for many years, but the catapult is violent in action, and a trolley running on rails and driven by an electric motor or by rockets may be found to give a better assisted take-off. It is curious how such ideas—which occurred to many pioneers of flight—are being revived to meet modern problems. Perhaps the most fascinating method that has been tried to assist take-offs is to mount a small, heavily loaded, high-speed aeroplane on a more lightly loaded but large high-powered aeroplane or flying boat. The larger aircraft takes off with the small one on its back or under its belly, and when reasonable speed has been attained, the small aeroplane is released to go on its high-speed journey while the mother craft returns to base. It is in this way that the X15, the American aircraft that holds the world speed record of 4,534 m.p.h. (1967), has been launched from a B52 Bomber at 45,000 ft. Another practical possibility is re-fuelling in flight, the aircraft taking off with a light load of fuel.

In the meantime, the only satisfactory way of dealing with the take-off of aircraft of high wing loading is to employ variable-camber flaps. By taking off with the flaps partially lowered the wing is converted into a high-lift wing, and the taking-off speed is comparatively low. Once in the air the flaps can be raised to the high-speed position. Unfortunately, however, flaps are not yet used to their full advantage for taking-off purposes, the reason being that the extra drag necessitates a longer run which may cancel out the good effects of a lower taking-off speed. The best way would be to have a flap which was lowered gradually during the taking-off run—but this is another complication.

Finally there are all the S.T.O.L. and V.T.O.L. types which are designed to give slow and vertical take-offs, just as they do slow and vertical landings.

74. Aerobatics

In the preceding sections we have considered various conditions of flight which are experienced by all aeroplanes. Whether designed for military or commercial use, for speed or for weight-carrying, whatever their type or size, aeroplanes must be capable of taking off, climbing, flying level, turning, gliding and landing. Beyond this certain types need not go; but others, such as fighting aircraft, must be capable of those much more violent manoeuvres which are usually classed as *aerobatics*. Under this heading may be grouped looping, spinning, rolling, flying upside down, steep side-slipping, violent stalling, etc.

The path followed by the aeroplane during such manoeuvres is very complicated, and it is not possible to reduce these conditions of flight to any simple theoretical analysis such as

AEROBATICS

we have attempted with the more normal conditions. Let us be content with a few interesting notes about the most important manoeuvres.

From the structural point of view aerobatics are of extreme importance: an aeroplane may be strong enough for straight and level flight, yet totally unsafe when flying inverted or in a nose-dive. Instruments called *accelerometers* are sometimes used to record the changes of load during the various manoeuvres, and the records obtained from these instruments have given us more valuable information than any abstruse theoretical calculations could ever do. The results may be summed up by saying that during the standard recognized manoeuvres, when performed by a capable pilot, the loads in the structure may be increased to four or five times those of straight and level flight. Occasionally the loads may be reduced, but this is not serious so long as they are not reversed, as in inverted flight. Notice that the figures given apply only to a capable pilot; a foolish and clumsy pilot can strain any aeroplane to its breaking-point. *Looping the loop* was probably the first recognized aerobatic manoeuvre, and remains to this day a standard feature of any aerobatic display.

A well-executed loop is rather like swinging a bucket of water round in such a way that no water falls out when the bucket is upside down. The pilot, like the water, will be sitting on his seat at the top of the loop without any danger of falling out; unlike the water, he will probably be strapped in, but there will be no load on the straps. Just as the pilot is in a normal condition—although not a normal position—so is the structure of the aeroplane, e.g. the load on the wings is in the usual direction (relative to the aeroplane) and has not been reversed as it is in real inverted flight. In short, although the aeroplane is upside down at the top of a loop, the conditions are those of normal flight and not of inverted flight. All this applies to a good loop; in a bad loop—well, anything may

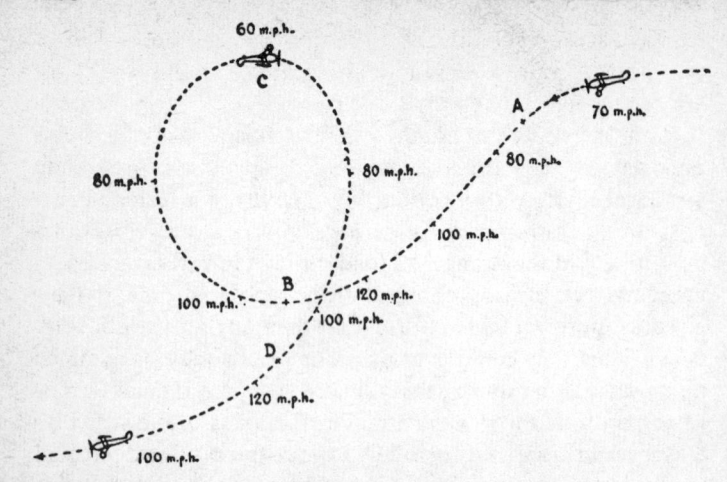

Fig. 94. **A loop**

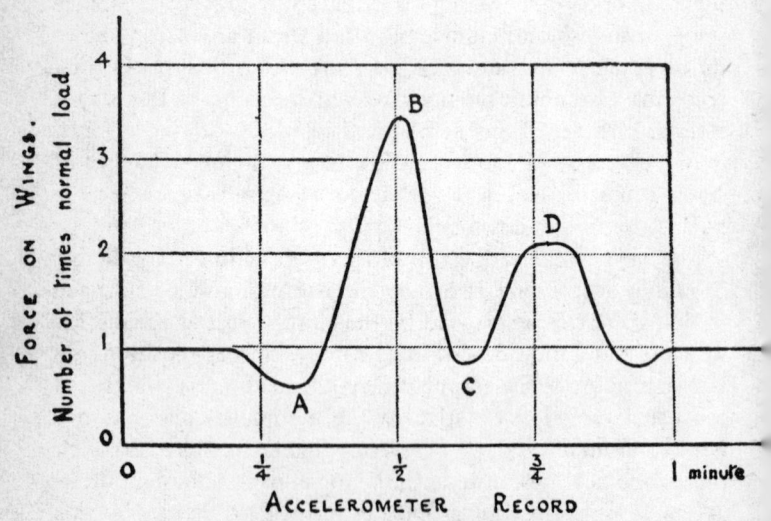

Fig. 95. **Record of loads during a loop**

AEROBATICS

happen. Figs. 94 and 95 show the approximate changes of speed and load during an ordinary loop on a slow type of aeroplane. Notice that the path travelled is by no means a circle.

The history of the *spin* is very different from that of the loop; for many years spinning was one of the unsolved problems of aviation. Aeroplanes would commence to spin for no apparent reason; some of them would not come out of a spin, whatever the pilot did with his controls, while others would come out of it if he did anything—or nothing. Some would spin fast, others slowly; some one way, some the other; some preferred a steep "spinning nose-dive," others a "flat spin." But the worst of it all was that no one seemed to know very much about it. Pupils were taught that if they were got into a spin it was not much use doing anything—they might try putting the controls neutral, but no guarantee was given that it would have any effect. It is no wonder that for a long time there stood to the discredit of the spin the majority of all aeroplane accidents. The pilot of today may be surprised to hear of this; to him the spin is merely an aerobatic, amusing rather than pleasant, a very effective means of impressing—and shaking up—a passenger unaccustomed to flying, but not dangerous. It is well that such pilots should realize that only after long and careful experiment and research has most of the danger of the spin been removed, and that the solution to the problem became possible only when we learnt what a spin was.

What, then, is a spin? It is an automatic rotation (sometimes called *auto-rotation*) of the aeroplane, set up as follows. The aeroplane is at, or near, the stalling angle, when for some reason one of the wings begins to drop. When it does so a wind will come up to meet it, and this, combining with the usual wind from the front, will result in an increase in the angle of attack on that wing. Now, at ordinary angles of attack, an increase of angle will cause an increase of lift, which will restore the

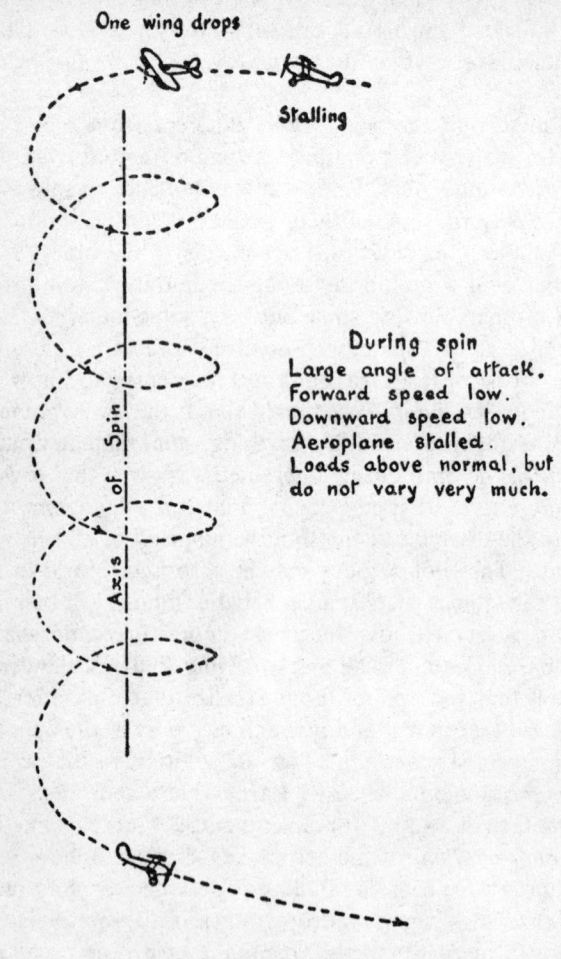

Fig. 96. A spin

AEROBATICS

wing to its normal position; but being at the stalling angle, *the lift on this wing will decrease*, and thus the wing which is already dropping will lose lift and will *automatically* continue to drop. Meanwhile the nose will drop, owing to the loss of lift, and auto-rotation, or spinning, will set in (Fig. 96).

The attitude of the aeroplane may vary tremendously, the the path is a very steep spiral, the forward and downward speeds are both comparatively low; but what is most important is that *the wings are stalled*, and therefore the action of the controls may be feeble, completely ineffective, or even have a reverse effect. This explains the original puzzle—we did not know that the aeroplane was stalled, and even if we had done, we would not have known how to gain control when stalled. It also explains why devices which prevent or postpone stalling and which give better control at the stall, such as automatic slots, also tend to prevent spinning or, at least, to give control over the spin. The size and position of all the tail surfaces, such as fin, rudder and tail plane, also influence the tendency of the aeroplane to spin.

The reader may have noticed that we have given both to stalling and to spinning the doubtful credit of being the cause of the majority of aeroplane accidents. The reason for this dual responsibility is that the accidental stall so often leads to a spin; if there had been no stall there would have been no spin, and moreover a stall that was not followed by a spin would often not prove dangerous. Thus it is difficult to say which is the real culprit—stall or spin.

To the uninitiated, a *roll* (Fig. 97) appears to be very similar to a spin, except that the axis of rotation is horizontal instead of vertical. It is true that the path followed during a roll is in the nature of a horizontal corkscrew, just as the path during a spin is a vertical corkscrew; but though the paths of travel are similar in the two manoeuvres, the principles involved are essentially different. The spin is an automatic and almost uncontrollable condition of flight, the nature and speed of the rotation being properties of the particular aeroplane, and the pilot cannot vary them to any extent. A roll, on the other hand, is a manoeuvre commenced and controlled throughout by the pilot. It is not likely to occur either accidentally or automatically, and the pilot is able to vary the speed of rolling.

But there are rolls and rolls; in a true slow roll (a rare feat) the aeroplane more or less turns around its longitudinal axis

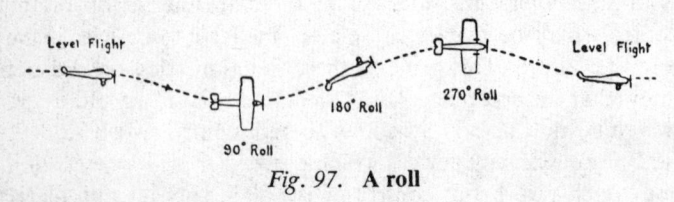

Fig. 97. **A roll**

and when it is inverted it is really upside down in that things tend to fall out; in a barrel roll, on the other hand, the path, as the name implies, is on a spiral or quite large diameter as if on the outside of a barrel, and when, half-way round, the aeroplane is inverted the loads are not—as at the top of a good loop things do not tend to fall out. Another extreme is the so-called *flick roll*, which really does correspond to a horizontal spin; the aeroplane is stalled at the start of the manoeuvre, which then occurs automatically.

A deliberate *side-slip* (Fig. 98) used to be a valuable manoeuvre for losing height or correcting drift in old-fashioned types of aircraft, especially when approaching a landing-ground. But today a real side-slip is hardly ever used, and would certainly not be classed as an aerobatic manoeuvre. A steep side-slip was never easy to perform and is virtually impossible in many modern types. During such a side-slip the wind strikes all the side or keel surfaces of the aeroplane, e.g. fuselage, fin, rudder. Owing to inherent directional stability, the turning effect of the forces on the rear keel surfaces will tend to cause the nose of the aircraft to drop, and the pilot finds it difficult or even impossible to maintain a steady side-slip. He must also take care to maintain sufficient forward flying speed; otherwise he will lose the effect of his controls.

In flying parlance an attempt is often made to distinguish between side-slipping and skidding, though both are really the same thing. An aircraft is said to side-slip if, when turning, the bank is too great for the speed; it is said to skid when the

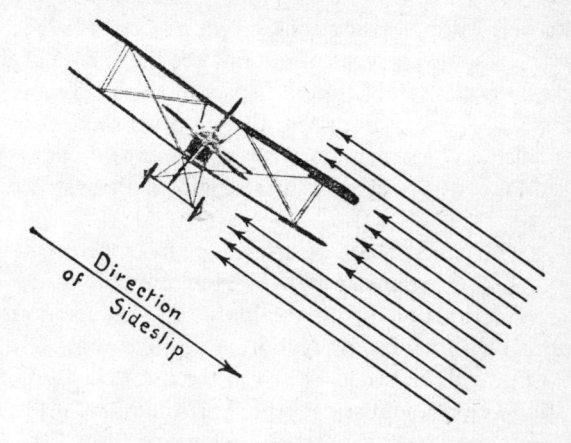

Fig. 98. **An old-fashioned manoeuvre in an old-fashioned aeroplane**

bank is too small. In this particular instance, the distinction seems clear, but it is not always so and even in this example is more apparent than real. Whether side-slipping or skidding, the fact that matters is that there is a sideways motion in addition to the forward motion, and the consequent side wind will strike the keel surfaces with all the usual effects.

Real *upside-down flight* (Plate 14) is much more rare than is commonly supposed. Quite apart from anything else, unless an aeroplane is specially equipped for the purpose, the engine will not run when in the inverted position. This is because the fuel supply will not function properly. Such an aircraft is perfectly capable of performing manoeuvres which have been mentioned already, even including those in which the aircraft may be temporarily inverted, such as a loop, and they may also *glide* in the inverted position. But continuous inverted flight without loss of height is another matter, and it must be admitted that it is a rather useless manoeuvre except for purposes of show and as a test of the pilot's ability to sit upside down for long periods without losing consciousness. Nevertheless, since fighting and sporting types of aircraft are liable to find themselves in almost any position, it is part of the conditions for a certificate of airworthiness in the aerobatic category that aircraft of the types mentioned should be capable of inverted flight so far, at any rate, as the strength of their structure is concerned.

Wings with concave under-surfaces are most inefficient when inverted, but a symmetrical or indeed any bi-convex wing section will function inverted almost as well as normally. Whatever the type of wing section, a positive angle of attack of the wings will be needed when in the inverted position. If the datum line of the fuselage were horizontal, i.e. if the aeroplane were in a kind of inverted "rigging position," the angle would be such that the lift on the wings would act downwards. So to provide an angle of attack, the aircraft must fly in a very much tail-down position (Fig. 99); e.g. if the rigger's angle of incidence is 4°, then in order to obtain an angle of attack of 4° when the aircraft is flying in an inverted position the datum line of the fuselage must be at 8° to the horizontal (tail down). The drag will thus be greater and extra power will be needed, the wings will be inefficient, and thus, to prevent loss of height, more speed will be needed to preserve the lift—in fact nothing is very satisfactory, but who cares? No one wants to fly very far upside down. As in a nose-dive, it is the structural problem

AEROBATICS

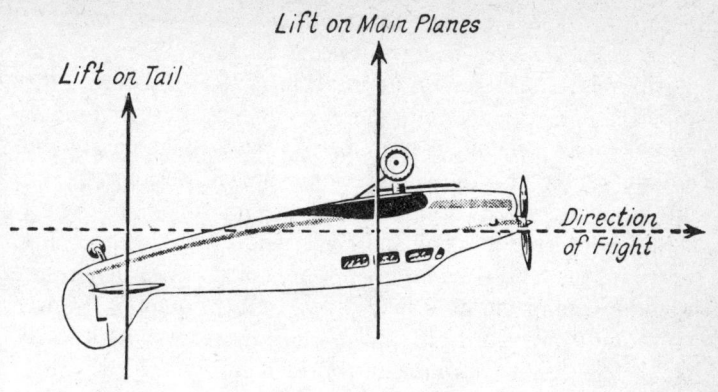

Fig. 99. **Inverted flight**

which provides most interest, because the loads are all in unusual directions.

We cannot pretend to have considered all the manoeuvres of which an aeroplane is capable. Double bunts, inverted spins, half-loops followed by half-rolls, half-rolls followed by half-loops, zooms, climbing turns, and so on—all these and many more form part of the pilot's aerobatic repertoire. Most of them are variations of those already given; they all affect the loading of the structure, and all have been investigated with accelerometers which record the increases and decreases of loads during flight. The records from these instruments will satisfy the structural designer that his framework is safe whatever mad things the pilot may do, and so we may leave it to the pilot who has discovered these manoeuvres to perform them to his heart's content. They improve his skill, they shake up his liver, and they may be useful in a fight—at any rate they are much easier to perform in the air than they are to figure out on paper, so we will say no more about them.

Apart from set manoeuvres and aerobatics, an aeroplane may suffer considerably from the effects of turbulence in bumpy weather. Anyone who has witnessed the havoc wrought by a rough sea, even if only viewed from the promenade of a seaside resort, can realize how a ship is at the mercy of the waves; but very few who have not experienced actual flight on a bumpy day have any conception of what a rough passage by air can be like. Gusty winds, bumps and air pockets, due to up currents and down currents, all are invisible, and even a skilled pilot cannot anticipate them. The aircraft without any warning may be carried bodily upwards or downwards, or one wing, the tail or the nose may drop. In fact it may be thrown into almost any position, and it will only recover its normal flight if the pilot uses his control, or if it is provided with a considerable degree of inherent or automatic stability. The continual use of controls for this purpose becomes very tiring to the pilot on a long flight in bad weather, and for this reason commercial aircraft, large bombers, and other machines required for long flights usually possess a high degree of natural or inherent stability, and are controlled for most of the flight by automatic pilots of various kinds. Bad weather in this sense does not necessarily imply either rain or wind, though thunderclouds are particularly bad offenders. The worst conditions are often found under a hot sun, with intermittent clouds, uneven ground conditions, and gusty, changeable winds.

75. The Propeller

We cannot leave our subject without saying a few words about the propeller which, in many types of aircraft, provides the thrust which in turn provides the speed which provides the lift that enables us to fly. It is true that if we were to follow this argument of the "house that Jack built" to its logical conclusion we ought to include the engine and the fuel—and even the sun, which is the original source of all our energy and without

THE PROPELLER

which flight would be impossible. But all this would be leading us into completely new subjects, whereas the theory of the propeller is simply an extension of the theory of flight.

The propeller blade is, in fact, nothing more or less than an aerofoil section or, rather, a series of aerofoil sections. When the propeller revolves, each of these sections strikes the air at a small angle of attack—just like a wing—thus causing a "lift" and a "drag." The lift, in this case, is forward; but it is a lift none the less, because it is "that part of the force (on the propeller blade) which is at right-angles to the direction of motion" (of the propeller blade). The drag is now sideways, but it is parallel with the direction of motion, and it opposes the motion. So much for similarity; the difference is that the "direction of motion" of some particular section of a propeller blade is a circular path instead of a straight line, or, to be more correct, it is a *spiral* path, because it is not only revolving but also travelling bodily forwards at the speed of the aeroplane. It is really very like a screw and is sometimes called an *airscrew*, but there is one important difference in that it cannot get the same grip as a screw can when revolving into something more solid than air. There is always a certain amount of *slip*.

We talk about the *pitch* of an ordinary screw, and the pitch of an airscrew is a similar idea—it is the distance that it would travel forward in one revolution if there were no slip. The distance it actually travels is less than this, and is called the *advance per revolution*. The difference between the pitch and the advance per revolution is the slip. The pitch may be several feet, being usually at least as great as the diameter, and it is largest in the fastest types of aircraft, since otherwise it would not be able to travel far enough in one revolution to keep up with the aeroplane which it is trying to pull along. Actually it must try to go *farther* than the aeroplane does, so as to get a grip on the air and exert thrust. The reader may suggest that if the revolutions of the engine were increased as the forward speed of the aeroplane is increased, there would be no need to increase the pitch. This is quite true, up to a limit; and engine designers like to develop their power by increasing their engine speed—this applies particularly to turbine engines.

But the limit is caused by the speed of the propeller blades, especially near the tip. When a propeller of 10 ft diameter is revolving at 2,500 r.p.m., the tip is travelling at a velocity higher than the speed of sound (1,100 ft/sec)—and in later sections of the book we shall see what that means. For this reason reduction gears were introduced so that the engine, which can develop most power at high speed, runs faster than the propeller, which is more efficient at lower speed.

The lift of each portion of the propeller blade, when added up, produces the *thrust*.[1] The drag of each portion has a turning effect, or moment, tending to stop the rotation, and when all these turning effects are added up they produce the *torque*,[1] which opposes and balances the total turning effect produced by the engine. This propeller torque will tend to turn the whole aeroplane in the opposite direction to the rotation of the propeller, and steps are usually taken in the design to prevent this.

Airscrews which pull from the front of the engine, causing tension in the airscrew shaft, are called *tractors*. Those which push from behind are called *pushers*—or perhaps we should use the past tense because the true pusher type of propeller seems to have disappeared from aircraft, which may seem curious in view of its universal use for the propulsion of ships. The term *propeller* is in everyday use as a substitute for the

---

[1] Strictly speaking, it is not quite true to say that the lift produces the thrust, and the drag the torque. The lift is the force at right-angles to the direction of motion of the blade, which is a *spiral* path. The *main* part of this lift, i.e. the forward part, is thrust, but the lift also contributes towards the torque. Similarly the drag of the blade, while mainly supplying the torque, detracts slightly from the thrust. The point is not very important so far as a book of this sort is concerned.

THE PROPELLER

word airscrew, but some technical people claim that it implies pushing and should only be used for pushers; others say that it means that the *air* is pushed, and that it is equally suitable for tractors or pushers. There is no point in our trying to resolve the controversy.

The *efficiency* of a propeller, like the efficiency of anything else, is the power that it gives out to the aeroplane compared with the power it takes in from the engine. Contrary to what

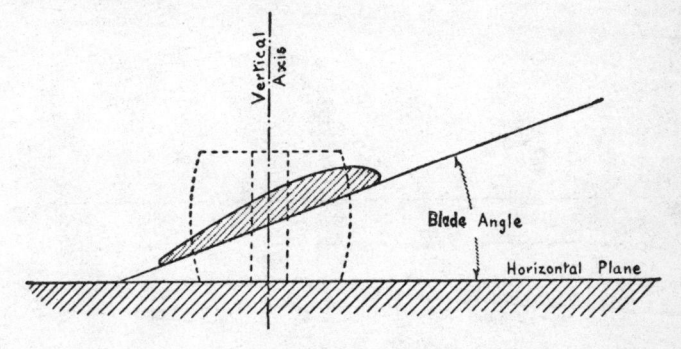

Fig. 100. **Blade angle**

might be expected, the efficiency of a propeller, when it is working at its best, is quite high compared with the efficiencies to which engineers become accustomed when dealing with other machines or engines. After all, the propeller is merely a machine; it converts the rotary power provided by the engine into the straightforward pull or push which we call thrust. Its efficiency may be as high as 80 per cent.

The angle at which a propeller blade is set (Fig. 100) is called the *blade angle*, or *pitch angle*. If the blade is examined carefully, it will be noticed that this angle is large near the boss, or centre of the propeller, and decreases gradually towards the tip. This is because, although all parts of the propeller must go forward the same distance in one revolution, the parts near the tip travel faster and farther in their circular motion than those near the boss. The portions of the blade near the tip may be compared to men climbing a gradual slope to reach the top of a hill, while the portions near the boss are like men climbing the same hill on a steeper path. If both are to reach the summit at the same time, those on the gradual slope will travel farther and faster (Fig. 101).

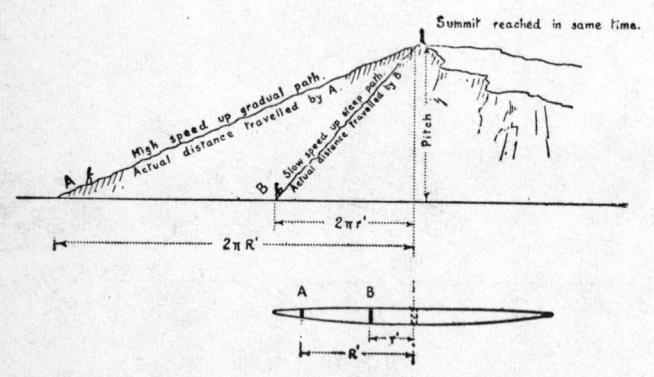

Fig. 101. **Analogy between the paths travelled by different sections of a propeller blade and paths of varying steepness to the summit of a hill**

It has been stated that the propeller blade *strikes the air at a small angle*, like the aerofoil. But the pitch angle of the propeller, even at the tip, where it is smallest, may be as much as 20° or 30°. In explaining this, the different gradients up the hill will help us again (Fig. 101), because these represent the paths travelled by the different parts of the blade. It is true that they have been straightened out into straight lines instead of spiral curves, but that is only to make them simpler to understand. Now, actually, any particular part of the blade is set at the angle required for the slope on which that part must

THE PROPELLER

travel *plus the small angle of attack* needed to give the thrust. You see the path of an ordinary aerofoil in normal flight is horizontal, or 0°, and the wings are set at the small angle of attack of say 2° or 3° to the horizontal, so it is much the same idea.

Since the pitch of a propeller depends on the forward speed and the revolutions per minute, the efficiency can only be at its best when the forward speed and the revolutions per minute correspond. This is a disadvantage in modern aeroplanes because every effort is being made to extend the speed range. For this reason much energy was devoted to solving the problem of producing a propeller with variable or controllable pitch, so that its blades could be rotated during flight to the best angle for the particular air speed and engine speed. Such a propeller has the additional advantage that the pitch can be altered to suit the changing density of the air with altitude, a higher pitch being needed to maintain a grip on the thinner air. Prejudice, mechanical difficulties, and the extra weight and complication involved all tended to retard the progress of this invention; but in spite of them all, its advantages were recognized, and the technical problems were solved. One consideration caused people to think of some kind of controllable pitch as a necessity rather than a luxury. The high maximum speed of modern aeroplanes means that the pitch of propellers must be high, so that for each revolution they can travel forward a long distance. High pitch means that the blades are set at a large blade angle. When travelling forward at the speed for which they are designed the blades will, of course, strike the air at a small angle. But when the aircraft is at rest on the ground, or is starting to move slowly over the ground for take-off purposes, the blades will meet the air at such a large angle (perhaps 70° or more) that they will be completely "stalled," and thus the lift or thrust will be very small and there will be difficulty in taking off.

It was for this reason that the two-pitch propeller became popular. Without giving all the advantages of real variable pitch, this propeller could be set in one of two positions, thus giving a fine pitch for taking off and a coarse pitch for high-speed flight. An improvement on the two-pitch type was the constant-speed propeller. In this, the blade angles are automatically changed during flight in such a way as to absorb the power of the engine while always allowing it to run at the same speed. This type of propeller is now in common use, and variable pitch has been even further extended to include *feathering* and *braking*. Feathering means turning the blades so that, when the propeller is stopped, they offer the least resistance; it is used in cases of engine failure. A propeller can be used for braking in two ways: first, by turning the blades so that they are at right-angles to the direction of motion and thus offer the maximum resistance—the exact opposite to feathering, in fact; secondly, by turning the blades even farther and then opening up the engine so that the propeller gives a thrust in the reverse direction—this is called *reversible pitch*.

The propeller must absorb the power given to it by the engine; otherwise it will simply race and have no effect. This means that it must have sufficient blade area, and that is why with increasing power of engines there was a tendency for the number of blades to increase from two to three, four, and even five. But there is a limit to the number of blades that can be fitted into one hub—and five is about the limit—and so we came to contra-rotating propellers, two propellers one in front of the other and rotating in opposite directions. These were advocated for a long time—because of their other advantages—but only came into their own over this problem of absorbing the power.

Another way of increasing the blade area so as to absorb more power is to increase the length of the blades, i.e. the

THE PROPELLER

diameter of the propeller, or their chords. Now the efficiency of a propeller, like that of a wing, is greater when the blades have a high aspect ratio, that is to say when the propeller has a large diameter and the blades have narrow chords. But alas, large diameters mean high tip speeds, greater stresses, and less strength to carry them, and—the final deciding factor—difficulty in clearing the ground without making the undercarriage unduly high. So in modern propellers efficiency has

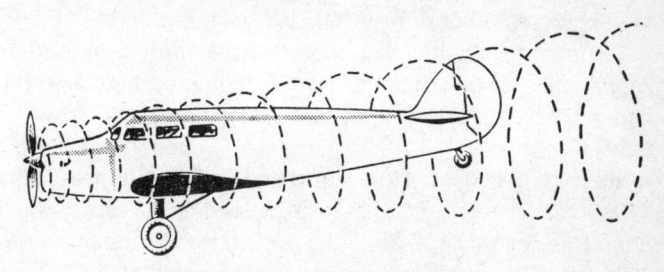

Fig. 102. **The slipstream**

to be sacrificed, diameter has to be restricted, and the chords of the blades have to be widened, sometimes right up to the tip, so as to absorb the ever-increasing power (Plate 57).

Just as the wing provides lift by pushing the air downwards, so the propeller provides thrust by pushing the air backwards. The backward flow of air is called the *slipstream* (Fig. 102). But the rotating blade of the propeller also pushes the air sideways, causing the slipstream to rotate. The diameter of this rotating high-speed column of air remains very nearly the same as that of the propeller, at any rate for the short distance it travels over the aeroplane itself. One cannot see the wake left behind by the aeroplane as one can when standing on the stern of a ship, but if one crosses this wake in another aeroplane one soon realizes its existence.

The slipstream has most unpleasant effects on the aircraft. Owing to its velocity being higher than that of the aircraft itself, all parts in its wake will have extra resistance, and in the normal type of single-engined aircraft these parts will include fuselage, tail unit and parts of the undercarriage and centre section—in fact a very large proportion of those parts which cause parasite drag. Furthermore, the pulsating, unsteady high speed flow which is usually set up causes wear and tear of the skin or covering, and draughts and noise in cockpits and cabins. Another difficulty is that, owing to its rotation in one particular direction, the effects of the slipstream are not symmetrical: it will strike one side of certain parts such as the fin and tend to rotate the machine or make it yaw from its course.

But one good effect of the slipstream is that it improves the rudder and elevator control at low speeds, and especially when taxying; the extra speed of air flow which the slipstream causes over these control surfaces may be used most effectively.

In the pusher type of aeroplane we were saved from the ravages of slipstream and flying became quieter and in every way more pleasant. However, certain snags cancelled out these benefits. The weight was too far back and balance and stability were difficult to obtain; it was also impossible to obtain a satisfactory arrangement of the tail unit, and the propeller could not easily be kept clear of the ground for landing purposes. But whatever its merits or demerits, the true pusher aeroplane—such as the original Wright machine—is practically non-existent today. Or is it?—aircraft driven by jets or rockets, especially when these are at the rear (Plates 25, 27 and 29), are surely pushers, and they certainly save us from most of the slipstream problems though they add a new one in that the slipstream is hot, and so must not be allowed to strike parts of the aircraft at all, and may even do damage to aerodrome surfaces.

76. Multi-Engined Aeroplanes

The reader may feel that throughout most of this book we have assumed that we are dealing with single-engined aircraft. If this is so, it has only been to establish principles. In most of the problems we have dealt with there is no essential difference between aeroplanes with one engine and those with two or more. However the engines are arranged, the combined thrusts will always give a resultant thrust which corresponds to that of a single and more powerful engine. This does not mean that there are no advantages in multi-engined aircraft; advantages there undoubtedly are, but they are advantages in safety and reliability rather than in the theory or principles of flight with which we have been concerned.

Provided that flight can be maintained *with one engine*, the gain in reliability by using two engines is obvious; unfortunately there were at one time twin-engined aircraft which could *not* fly on one engine, and in such cases an extra engine is a liability and not a source of reliability. Many modern types have four or more engines, they can be flown very satisfactorily on any three, and can maintain level flight on any two; the reliability thus gained is of inestimable value on long flights across the sea or over difficult country.

The problem of making a twin-engined aircraft capable of flight on one engine is not a simple one. It is not just a question of lack of power (although that may be serious enough), but one has to counteract the tremendous turning effect of the remaining engine. This needs powerful rudder control, and, to be fully effective, the rudder in propeller-driven aircraft must be in the slipstream of the remaining propeller. Since either engine may fail, there should be a rudder in the slipstream of each propeller. A bias gear may be introduced so that the rudder can be set to give a permanent turning effect in one direction without the pilot having to exert continuous pressure on the rudder bar.

The turning effect in multi-engined aircraft resulting from the stoppage of one or more engines will clearly be more serious if the engines are far apart. This is another disadvantage of propellers having large diameters, but it is also the clue to one of the great advantages of multi-engined jet propulsion as against multi-engined propeller propulsion. The difference is clearly illustrated in the examples of multi-engined aeroplanes of both types which will be found in the Plates, and in which it will be noticed how close the nacelles of the jet engines can be, both to the body of the aircraft and to each other. But this is not always so (Plate 26)—for there are structural and other advantages in spreading the load along the wing. A modern trend in the design of multi-engined jet aircraft is to locate the engine nacelles at the rear of the fuselage and on either side of it (Plates 25 and 53)—who said that the true pusher had disappeared?

77. Flying Faults

Those readers who are unaccustomed to flying may imagine that an aeroplane, being a mechanical device, ought to behave with the precision of a machine. Perhaps it ought, but it certainly does not. Aeroplanes seem to possess faults, tempers as bad and habits as vicious as those of any human being. Such faults may never be cured and the type in consequence never come into general use. On the other hand, some slight modification has often been found to turn failure into success; miracles have been performed by the substitution of a four-bladed propeller for a two-bladed one, by alterations of fin or tail plane, by change of engine, and so on. The reasons for the remedy are often as obscure as the reasons for the fault, and, what is more, remedies which produce

FLYING FAULTS

wonders in one type may be completely ineffective in another.

Apart from these mysterious faults—accidents of design we might almost call them—aeroplanes develop from time to time a tendency to fly incorrectly if left to themselves; to fly say with one wing lower than the other, or to tend to turn to right or left, or to be nose- or tail-heavy.

Now, our attitudes towards these faults has changed completely in the last thirty years or so—indeed, since this book was first published. In the old days, and especially in the case of biplanes, aeroplanes were more adjustable than they are now, and if for any reason they got out of true, the rigger was able to adjust them by the simple process of slackening and tightening wires.

But whether we like it or not, those days are gone, and with them the rigger himself; very little, if any, adjustment can be made to the "rigging" of the modern aeroplane on the ground. But another significant change has taken place too, and it is just as well that it has, because with all the rigidity of the modern aeroplane, and all the precision that is put into its manufacture, it is still apt to develop flying faults, and it would be most unfortunate if there were no means of correcting them. But now they are so easy to correct by trimming adjustments in the air that the pilot hardly recognizes them as flying faults at all.

These changes represent progress, but they have had one unfortunate result. The tracking down of flying faults, the diagnosis as it were, and the application of the remedy, provided the essential link between the theory of flight and the practice of flight; there was no better way for practical men like the rigger and the pilot to learn the principles of flight than to try to puzzle out what was wrong and how to put it right. Nowadays if the aeroplane tends to fly with one wing low the pilot just turns a knob; he may or may not know what happens when he turns the knob, but it is almost certain that he doesn't know why he had to turn it, i.e. what was making the aeroplane fly with one wing low.

So, to help the reader to understand, I am going to ask him to use his imagination and to put the clock back to the days when flying faults had to be diagnosed and corrected by the rigger on the ground; or, if he prefers it, to think of the faults as applying to a model aeroplane which even today can usually be adjusted sufficiently to correct the faults.

Suppose, for instance, that an aeroplane *tended to fly with one of its wings lower than the other* unless the pilot pushed the control column over to the opposite side to raise the lower wing. Perhaps it would be better to say that the left wing was heavy, and that the pilot had to lift it in order to fly on an even keel. If this fault develops in a model it will actually fly with one wing low—with consequences that will be mentioned later—unless it is equipped with some automatic stabilizing device, or can be controlled by radio from the ground.

Now, ideally, the rigger or the owner of the model should first discover the *cause* of this or any other fault. But in practice this has its disadvantages; for aeroplanes, like human beings, seem to have peculiar "kinks", and of two apparently identical aeroplanes, one may fly with one wing heavy and the other will not. Both may be checked all over, all dimensions may be found correct, and yet the one will continue in its bad behaviour while the other flies perfectly. In such circumstances, if some adjustment could be made to the faulty aircraft so as to make it fly correctly, this should be done *even if the rigging dimensions were incorrect after the adjustment*. For what really matters is whether the aeroplane can be made to fly correctly. On the old biplanes the accepted method of achieving this result was to increase the rigger's angle of incidence on the wing which tended to fly low, to decrease it on the opposite wing, or to do a little of each. It might be thought that it

FLYING FAULTS

would be sufficient to rig the ailerons so that when the control column was central the left aileron (if flying left wing heavy) was slightly depressed and the right aileron neutral or slightly raised. This might have some effect, but the result is likely to be disappointing; the ailerons would tend to find their own level, the control column would move slightly over to the left, and the aircraft would still tend to fly left wing low. Those who are not pilots themselves often seem to imagine that a pilot tries to fly straight and level by placing his control column and rudder bar *in central positions*. But pilots fly chiefly by *feel*; if, with hands off, the aircraft flies on a level keel, the pilot does not bother if the control column is one inch to the left, nor will he report the machine as flying right wing low. *It would fly right wing low if he put the control column central, but he does not; he lets it stay where it wants to be.* What the pilot notices is when he has to *push* the control column to right or left in order to make the aeroplane fly on an even keel. If we can make the left-hand aileron tend of its own accord to keep in a lower position than the right-hand aileron, then the required result will be achieved. In the "good old days" this was sometimes done by fitting elastic to the control column from the side of the cock-pit, and tightening the elastic according to the amount of bias required. Nowadays this is done by use of the trimming tabs which were described in Section 57.

While we have pointed out that the chief aim was to make the aeroplane fly correctly—provided, of course, that nothing was done to cause danger or spoil performance—it is interesting to consider what might have been the possible *causes* of a machine flying with one of its wings low. For example, suppose that it was the left wing. The torque of a propeller revolving clockwise (as seen from pilot's cockpit) would tend to make the aeroplane revolve in the opposite direction, i.e. to drop the left wing. Probably the designer would have allowed for this somewhere in his calculations; if he had not, that might have been the cause.

Too much incidence on the right-hand wing would cause extra lift on that wing, too little incidence on the left-hand wing would cause a loss of lift. Either or both of these might make the left wing heavy. If the camber on the left wing was distorted, or if there was anything to decrease its lift, it would tend to fly low.

If there were some large weight such as a bomb or a full petrol tank on the left wing without a corresponding weight on the other wing, the left wing would tend to drop. This is rather obvious, and it is most likely that the cause of it would have been equally obvious.

Some books used to tell you that incorrect dihedral on one of the wings might be the cause. But suppose, for instance, there was 1° too much dihedral on the left wing—quite an appreciable error—the aeroplane would, in theory at any rate, adjust itself so that there was no side-slip, i.e. so that there was the same *effective* dihedral angle on both sides. The pilot would then be sitting on a seat inclined $\frac{1}{2}$° to the horizontal. No one who has ever flown can think of this as a very serious matter; incidentally, too, neither would the aircraft be flying with the left wing low, nor would the left wing be heavy.

An aeroplane too may be either *nose-* or *tail-heavy*—this time we use the more sensible terms, and this time we feel justified in using the present tense, because even the most modern of aeroplanes can exhibit this fault, and if the cause is in incorrect loading, it can be discovered and rectified. Before the days of adjustable tail planes and trimming tabs this was the most common of all flying faults. We do not hear so much of it nowadays, not because it does not occur, but because it is so easily remedied by the pilot during flight; all that is required is a slight adjustment on the elevator trimming tabs. By this method we are certainly applying the remedy without discovering the cause; so much so that there is a danger of a fault which is really due to a distorted structure or incorrect loading not being reported. It should be remembered that an aeroplane may be trimmed to fly hands off even when there is some serious fault in its rigging or loading; but it can never fly at its best in such a condition, and therefore nose- or tail-heaviness should always be tested by flying the aircraft with the trimming tabs in the correct position for normal flight.

Clearly the aeroplane will be nose-heavy if either the centre of gravity is too far forward or the centre of lift is too far back. It is very sensitive even to slight movements to these two big forces. Air-worthiness regulations lay down that the centre of gravity shall be within narrow limits. Very careful loading of aircraft is always necessary, and sometimes ballast weights must be carried so as to keep the centre of gravity within the required limits, in spite of the fact that weights may cause a falling-off in performance. For instance, in some two-seater aircraft, if no passenger is carried in one of the cockpits, specified ballast weights must be carried instead. *It is particularly important that the centre of gravity should not be too far back*, since this fault may make the aircraft dangerously unstable.

A tendency to turn to right or left is one of the most irritating of all flying faults. Except when flying in clouds or fog, a pilot is very soon aware of a rolling or pitching of the aircraft even without consulting his instruments; but unless he watches his instruments or the ground very carefully he may allow the machine to yaw several degrees off its course before he notices it. Another annoying feature of the fault was that the cause was difficult to detect, and in some instances there was no obvious remedy. If the tendency to turn off the course was sometimes to one side, sometimes to the other, then it was really a question of directional instability, and both pilot and rigger were more or less helpless; it was a matter for the designer to investigate and possibly to fit a larger fin. If, on the other hand, the tendency was always to turn in *one* direction, e.g. to the left, it was possible to apply some form of opposite rudder effect, and the universal modern method of doing this is by using trimming tabs on the rudder.

Before leaving this subject, we ought perhaps to mention that, in this case as in the others, any *obvious* causes of the trouble should be investigated first. For instance, any obstruction on the left-hand wing will cause extra drag on that wing and thus cause the aircraft to yaw to the left. So powerful is this rudder effect of extra drag on one wing that it has often been suggested that some means of increasing the drag on each wing as required would make a better "rudder" than the conventional design.

One of the most puzzling of all the faults to which an aeroplane is liable is that of excessive *vibration*. When we consider all the rotating and reciprocating parts in an elaborate aero engine, the revolving propellers and the comparative flexibility of all the parts of the structure, it is hardly surprising that vibrations should occur, and it needs some experience to detect the change, sometimes a gradual one, from the inevitable vibration which must always be present in such an elaborate construction to the vibration which indicates some fault in the mechanism. If this latter vibration is allowed to continue unchecked, the chances are that it will go from bad to worse and may very soon end in disaster. All cases of vibration should be investigated at once, and this differs from the other faults in that *the cause must be detected before the remedy can be applied*. The range of investigation is also larger; and it is by no means confined to the structure of the aeroplane. Many cases in practice have been wrongly diagnosed. A loose propeller, an irregularly running engine, or an overtight bracing wire, all may cause a vibration which is felt throughout the airframe. In the aeroplane structure itself, the danger of vibration is that it may lead to the type of flutter that has already been mentioned (Section 59) in connection with control.

We do not pretend to have considered all the faults that may occur in flight. It was stated at the outset that each particular aircraft may have its own individual faults, and this applies even more so to the separate types. Some are difficult to land, others easy; some will tend to spin when they should not, others will not spin when one wants them to do so; some may have faults of stability or control or be subject to little tricks to which the pilot must become accustomed; very often the designer is quite unaware of these features until the test pilot has put the aircraft through its paces, and by that time, unless the fault is really serious or dangerous, little can be done about it. Thus the type will go into service with all its individual characteristics, and pilots and all who work on it will have plenty of little things to grouse about; but after all—I was going to say "It is only human!"—well, it comes to much the same thing—nothing can be perfect, and perhaps it is just as well that aeroplanes are no exception to the rule.

78. Instruments

The dashboard of a modern motor car is an attractive enough sight to anyone who is fond of gadgets, but it is nothing in comparison with the cockpit of a modern aeroplane. In both the motor car and the aeroplane there is much significance in the word "modern." The early cars may have been fitted with a speedometer of doubtful accuracy and a clock which usually remained at the same hour for weeks on end, but some of the early aeroplanes had nothing at all in the way of instruments, and even when they did, the pupil pilot *was severely discouraged from taking any notice of them.* "Flying by instruments" was a term of reproach. The proper way to find your way about the country was to look at the ground (if you could see it); the proper way to know whether you were flying at the right speed, banking at the correct angle, and so on, was to "fly by instinct," whatever that might mean. No wonder the popular Press of those days called us "bird-men." Small wonder, perhaps, that they usually added the epithet "intrepid"! As for the engine, well, who cared what the oil pressure was? Usually there was none, because the oil was just splashed about all over the place, and we were supposed to be such mechanically minded bird-men that we could at once detect whether the engine was running at the correct speed.

Mind you, there was much sense in the ideas of those early days. For one thing, you had to learn to *fly*, it was not just a question of being clever enough to keep a lot of needles all pointing to their correct readings; and secondly, the instruments of those days were often so unreliable that it may have been wiser to trust to one's rather doubtful bird-like instinct. I often recall an early solo flight in which I was gliding in to land in an old Caudron; the machine was easy to fly, it was a glorious summer evening, and everything in the garden seemed lovely—so lovely in fact, and so calm was the air, that I was watching my air-speed indicator and trying to keep the needle steady at 50 m.p.h. When it went round to 52, I eased the stick back a little; one did not learn instrument flying in those days, but I did at least expect the needle to go back to 50. Naturally it was somewhat surprising, but not yet alarming, to see it move to 55. Ah well, the stick must come farther back, and so it did; but the needle did not go back, it went on to 60, then 65, 70, 80, 100. . . . No, I didn't wake up in hospital; it is difficult now to say at exactly what figure I did wake up; but when I did, and looked over the side, the nose of the machine was sticking up in the air, well stalled, and there was a not very inviting wood down below. It was the qualities of

THE AIR-SPEED INDICATOR

the old machine rather than any bird-like instinct that enabled me to land safely after that alarming experience, but I did at least feel that there was some sense in what my instructor was saying when he indulged in a long but very unchurchmanlike sermon, of which the text was "Thou shalt not fly by instruments."

But, happily or unhappily, those days are past; the modern pilot, though he may still try to develop his bird-man's instinct (which he now describes more vividly as *flying by the seat of his pants*) is taught to use and to trust his instruments, and at times he is entirely dependent upon them. He even does a course of instrument flying, in which he is shut up in a box not only during flight, but sometimes for taking off and landing as well; he is rated according to his ability and given cards of various colours as evidence. This may seem to be going a bit far, but after all, one cannot find one's way across the Atlantic by instinct; and fog, the greatest enemy of flying, defeats not only the most bird-like of all airmen, but even the birds themselves.

Thus it is that the aircraft of today is equipped with a great multitude of instruments, and that these have been so developed that they have now reached a very high degree of accuracy and reliability. To describe them all in a book of this kind is clearly out of the question—a mere catalogue would fill a whole paragraph—therefore we must content ourselves with mentioning a few interesting points about those instruments which most nearly concern the actual flight of the aeroplane.

79. The Air-Speed Indicator

Throughout the book we have talked of air speed, and we have repeatedly noticed the close connection with angle of attack. In taking off, climbing, straight and level flight, turning, gliding, and landing, there is a best speed for each, while for the purpose of flying from one place to another the navigator must know both the air speed of the aeroplane and the velocity of the wind. It is true that he would prefer to know the *ground* speed, but no instrument can be devised to measure this directly, and the pilot much prefers to know his air speed.

The usual type of air-speed indicator consists of a thin corrugated metal box very like that used in an aneroid barometer. At some convenient place on the aeroplane, where it

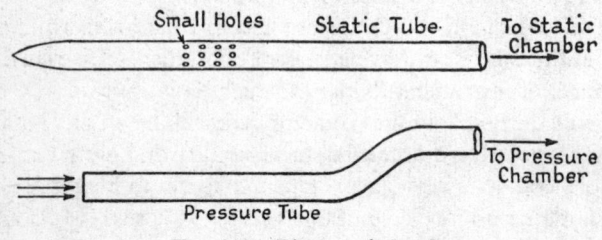

Fig. 103. **Pitot-static head**

will be exposed to the wind yet not affected by slipstream or other interference, is placed the *pitot-static head* (Fig. 103). This consists of two tubes, one of which has an open end facing the air flow—called a *pitot tube*. The other is closed at the end, but along the sides are several small holes which allow the atmospheric pressure to enter, and this tube is called the *static tube*. In modern types the two tubes are often combined into one, the static tube being concentric with the pitot tube, and outside it (Fig. 104). Sometimes the pressure near the pitot

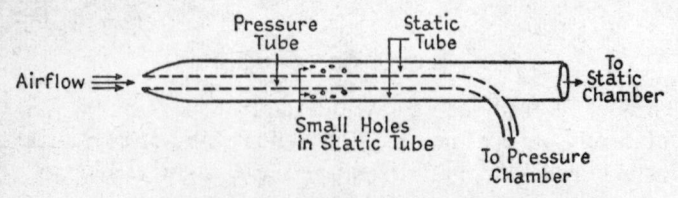

Fig. 104. **Concentric pitot-static tube**

THE AIR-SPEED INDICATOR

tube is by no means atmospheric, and the static pressure is taken from some other part of the aeroplane altogether. But wherever the pitot head, and the static vent, may be, metal tubing is used to communicate the pressures to the instrument in the pilot's cockpit, the pitot tube being connected to one side of the metal box and the static to the other. When the aeroplane is at rest relative to the air, the ordinary atmospheric pressure will be communicated by the tubes to both sides of the box and the instrument needle will be at "0," but when travelling through the air the pitot, or open, tube will record a higher pressure, depending on the air speed, while the static tube will still record the atmospheric pressure. The instrument then reads the difference between these two pressures which is automatically translated by the dial into miles per hour or knots.

The pressure on the pitot tube, just like all air resistances, will go up in proportion to the square of the speed, e.g. at twice the speed the pressure will be four times as much, and thus we can understand why the numbers round the dial of the instrument, 50, 60, 70 m.p.h., and so on, are not equally spaced.

When we fly higher, the density of the air will become less, and since the difference between the two pressures depends on the density as well as on the air speed, the indicator will read incorrectly. We call the speed recorded by the instrument the *indicated air speed*, and the real air speed the *true air speed*. The error is quite appreciable; for instance, when the indicator reads 100 m.p.h. at 30,000 ft, the true air speed is about 160 m.p.h., and at 40,000 ft a reading of 100 m.p.h. on the instrument means that we are really travelling at more than 200 m.p.h.

There is, however, rather an interesting point about this incorrect reading of the air-speed indicator at height. Just in the same proportion as the pressure on the metal box is reduced by the smaller air density, so is the lift on the wings correspondingly reduced, and thus a higher speed is necessary to support the aeroplane in flight. Therefore the stalling speed of the aeroplane will increase with height, but at this increased speed the air-speed indicator will continue to read, when the aeroplane is about to stall, the same stalling speed as when near the ground. The error, in other words, has a distinct advantage from the pilot's point of view in that, whatever the height, the aeroplane stalls at the same *indicated* speed. Other speeds of flight, such as the speed for maximum range, are affected in the same way.

True air speed can be measured by a system of rotating vanes or cups called an *anemometer*. This instrument is used at meteorological stations for measuring wind velocity, but it is not very satisfactory for use on aircraft. For navigational purposes elaborate instruments have been devised for measuring true speed, but they are outside the scope of this book.

80. The Altimeter

The word "altimeter" means "height measurer." Would that the instrument were true to its name! The so-called altimeter which is used in aeroplanes is nothing more or less than an aneroid barometer, such as is used to measure the pressure of the atmosphere for the purpose of forecasting the weather. The only real modification is that the dial is marked in thousands of feet instead of in inches or millimetres of mercury, and this makes it just about as capable of measuring the height as the barometer is of foretelling the weather. What it does do is to record the *pressure*. As we go up, the pressure goes down, because there is less weight of air on top of us; but unfortunately the rate at which the pressure goes down varies from day to day, depending chiefly on the temperature and other effects, which also vary from day to day. Thus it is impossible

THE ALTIMETER

to mark off or calibrate the scale of an altimeter so that each pressure corresponds to a definite height; the best that can be done is to assume some average set of conditions of temperature and pressure, to mark the scale of the instrument to suit these conditions, and then correct the readings for any large departure from such standard conditions.

This set of average conditions has been laid down, and, as mentioned in Section 4, is called the *International Standard Atmosphere* (Fig. 11). When an aeroplane makes a test flight, or some attempt on an altitude record, the height which counts is not the height reached according to the altimeter, nor is it the actual height above the ground; it is the height which we estimate it *would have reached* had the conditions of the atmosphere all the way up corresponded to those of the Standard Atmosphere. It is not a very satisfactory state of affairs, but we cannot do any better until we can devise an instrument which will really measure height, instead of just pressure.

Not only does an altimeter fail to record the correct height when flying, but it does not necessarily read zero when at sea level, since the atmospheric pressure varies considerably from time to time at the earth's surface. After all, that is how a barometer works, and the altimeter is only a barometer. For this reason, altimeters are fitted with an adjustment so that they can be made to read zero (or the height of the aerodrome) before starting on a flight. It does not by any means follow that they will read zero on return to earth. In a flight of a few hours there may be considerable change in atmospheric pressure, and there is also a certain amount of lag in the instrument. For these reasons it is very important when flying over high ground or mountainous districts in foggy weather not to put too much faith in the altimeter. Although this is usually impressed upon pilots, accidents have occurred from this cause.

Modern altimeters are very much more sensitive than the old types. Some of them have three hands, one making a complete revolution every 1,000 ft, the next one every 10,000 ft, and the third in 100,000 ft. There is hardly any lag in such an instrument; in fact, such sensitivity would be of no advantage if there were any serious lag. Another refinement is that, instead of turning the dial to set the zero, the pointers are moved, and when they read zero a little window at the bottom of the instrument gives the reading of the barometer. A great advantage of this method is that if one can find out, by radio or other means, the reading of the barometer at any aerodrome at which one wishes to land, one has only to set this reading on the altimeter and, whatever may be the altitude of the aerodrome, the hands will all point to zero or, by an alternative setting, to the correct height of the airfield, when the aircraft touches the ground. This is a great help in instrument flying.

But, however sensitive the barometric type of altimeter may be, it still cannot measure true height in the atmosphere, except under a very unlikely set of standardized conditions.

Is it possible, then, to measure the true height of an aircraft above sea level? In certain instances it can be done by taking three simultaneous sights from the ground, or by various radio and radar devices, or by some echo system such as is used for submarines.

But, for most purposes, the altimeter, the aneroid barometer, with all its faults still holds its own, and though we never know how high we are flying we can either assume ignorance and hope that the altimeter is right, or we can try to be very clever and work out how high we ought to be. Special "computers" are provided for this purpose.

81. Navigation Instruments

For any kind of cross-country flying the magnetic compass, until the introduction of radar in the Second World War, was

NAVIGATION INSTRUMENTS

as essential a part of the equipment of an aircraft as it is of a ship. It is true that in clear weather there are certain advantages in following by means of a map the landmarks on the ground rather than in trusting to the compass, so much so that at one time some pilots used to argue that a compass was unnecessary, and in any case could not be trusted. One point which used to be brought up against it was that when one entered a cloud or fog, i.e. just when the compass was needed, it used to go round and round. The reason it did so is now well known—it went round and round because the pilot was steering the machine round and round in circles, chasing his own tail, yet thinking all the time that he was going straight ahead. If you do not believe that this is possible you can never have been on a moor or a large, open field, or even on a wide road, in a thick fog.

But in these days pilots are better educated; they know that sooner or later they will have to fly through clouds or fog or over the sea, and that in such circumstances, if the aircraft is not fitted with radar devices, the compass is their only guide. There is, however, one respect in which navigation by compass can never be quite so satisfactory in the air as it is on the sea. In deciding on the correct course to steer one must make allowance for the wind, and the wind is a very uncertain and changeable quantity. It is true that the navigator by sea must allow for currents and tides, but usually these are smaller in comparison with the speed of the ship then the wind may be compared to the speed of the aircraft, and, what is much more important, the strengths of the currents and tides are usually very accurately known and can be listed in tables and on charts, whereas no one knows from one hour to another from what direction and with what velocity the wind may blow. For this reason, a wise pilot will whenever possible check up his compass by radio direction-finding, and by landmarks on the ground.

The aircraft compass has been so much modified from the ship's compass that it is scarcely recognizable as the same instrument. In comparison with the mariner's compass it must be lighter in weight, quicker to take up its new reading yet damped so that it does not oscillate too much; able to withstand very low temperatures at great heights; to put up with the vibration of the machine; and, perhaps most difficult of all, to continue its job as a compass when the aeroplane takes up attitudes such as no ship would ever experience.

In common with its opposite number, the aircraft compass must be corrected as far as possible for *deviation*, i.e. for errors produced by magnetic material in the aircraft. This correction is made, just as on a ship, by "swinging" the aircraft, i.e. turning it round on a compass swinging base on the ground so that it points towards the known compass points and correcting the reading by placing small correcting magnets in slots above or below the compass. The correction can never be complete, and a small table of "deviations" is usually put on the instrument board in the cockpit. Again, like the ship's compass, the aircraft instrument being what it is, i.e. merely a magnet, will point along the magnetic meridian and not the true meridian. This error is called *variation*, and its value varies at different parts of the earth's surface and also from year to year, but all this and many other interesting and important points in connection with the compass really come under the subject of aerial navigation and cannot be discussed in this book. The same applies to the multitude of electronic devices which have revolutionized the navigation of aircraft; these have enabled us to "see" our way in the dark, and in fog, and through the clouds, and navigation in all these conditions can be as easy nowadays as it was when we used to follow the railway lines on a clear day—as easy, that is, once we have learnt to operate and interpret the instruments correctly and provided, of course, that they don't go wrong.

82. Flight Instruments

After the stress that was laid in an earlier paragraph on the number and variety of instruments which are used on modern aircraft, it may seem rather strange to dismiss all the others in a couple of paragraphs. But the fact is just that there are so many that it would be unreasonable to attempt to describe them here. The three already mentioned—air-speed indicator, altimeter and compass—may perhaps be considered as those most directly concerned with the subject with which we have been dealing, the flight of the aeroplane as such. To these should perhaps be added the *turn and side-slip indicator*, in which the slip needle or ball has replaced the old cross-level, which was nothing more than a spirit-level with a very pronounced curve, placed laterally across the aeroplane. The needle or ball makes an excellent indicator of side-slip, as was explained in Section 70 when the turning of an aeroplane was being considered. It has sometimes been called a "bank indicator," but this is really a bad name for it, since the aeroplane may have its lateral axis inclined at a large angle, e.g. in a steep bank, yet the slip needle or ball will remain central—because there is no side-slip. It is a simple instrument, yet a very important one, and possibly the pilot's best guide as to whether he is flying correctly or not.

But it was the gyroscopic type of instrument which revolutionized instrument flying. A gyroscope (more familiar perhaps as a top—incidentally a very scientific toy) is a rapidly rotating wheel which possesses two main features. First, its axis of rotation tends to remain rigidly fixed in space; secondly, if an attempt is made to tilt its axis it will tend to rotate about a third axis. Consider, for instance, the front wheel of a bicycle; if it is tilted over so that its top goes to the right, the whole wheel will turn to the right (that is why it is possible to ride a bicycle without one's hands on the handlebars). This second property of a gyroscope is called *precession*.

An instrument panel in a modern aeroplane may contain at least three instruments which depend on gyroscopes. They are usually driven by suction from an engine-driven pump or from double venturi tubes exposed to the air stream, and may revolve at 10,000 r.p.m.

Perhaps the most striking of all such instruments is the *artificial horizon*, which shows the position of a small model aeroplane relative to a horizon marked on the instrument. If the nose of the real aeroplane goes down, the model goes below the horizon; if the nose goes up, the model moves above the horizon. If the aeroplane banks to right or left, so does the model. Even if the pilot cannot see the real horizon at all, if he is flying on the darkest of nights, or "under the hood," he can always tell the attitude of his aeroplane. Only those who have tried to fly "blind" can possibly conceive the value of such an instrument. It is worked by a gyroscope which is so mounted that its axis does not move even though the aeroplane (and with it the case of the instrument) may pitch or roll.

Simpler in principle, but no less useful in practice, is the *directional gyro*. This detects any turn of the aeroplane, just as the artificial horizon shows pitch or roll. It is very like a compass except that, instead of possessing the property of pointing towards the north, it will remain in any position in which the pilot likes to set it. Actually it is marked off in degrees just like a compass, and the pilot usually sets it to correspond to the compass course. The reader may well ask what its justification may be, seeing that it seems to act like a compass, though lacking the chief attribute of the latter. The answer is simple. The directional gyro responds more quickly to the slightest turn, it settles down at once after a turn, it is unaffected by accelerations and the various magnetic errors of the compass.

These two instruments together—the artificial horizon and the directional gyro—are the basis of "George," the robot or automatic pilot, which not only detects any tendency of the aeroplane to yaw, pitch or roll but, having done so, moves the controls until it is once more flying correctly. That sounds wonderful indeed; but it is no longer fantastic to imagine that in the future aeroplanes will be flying about, carrying and dropping bombs, and perhaps even fighting each other, without any pilots at all—indeed guided missiles are already doing just this.

The third gyroscopic instrument in common use is the *turn and side-slip indicator*, which has already been mentioned. The lower needle on this indicates the *rate of turn* and is worked by the precession of a gyroscope; the upper needle indicates *side-slip* and is worked by a pendulum.

There are not many other instruments concerned with the actual flight of the aeroplane. The air temperature is needed for various corrections to speed, height, and so on in record or test flights, and for this purpose an ordinary *thermometer* may be fitted on some exposed part. A *rate-of-climb indicator* or, to be more exact, an instrument which shows either rate of ascent or rate of descent, is usually fitted to modern aircraft, and, like so many of these modern luxury instruments, is of great value in instrument flying. A *machmeter*, which will be mentioned in the following paragraphs, is indispensable in high-speed aircraft.

Apart from the aircraft itself the engine or engines will need revolution indicators, oil-pressure gauges, oil-temperature gauges, air-pressure gauges, fuel-pressure gauges, boost gauges for superchargers, water thermometers for water-cooled engines, fuel flowmeters, fuel-contents gauges, and so on.

On the electrical side there may be anything varying from the simple switch used for the engine ignition to a complete system of lighting and heating, dynamos and motors, and full radio and radar installation with all its attendant instruments.

For high flying, oxygen apparatus must be installed, and this needs special instruments all to itself as does the pressurization of cabins.

Incidentally, we must not forget what is perhaps the most useful of all man-made instruments—the *clock* or *watch*. For any kind of serious flying it is indispensable.

83. High-Speed Flight

But the time has come to fulfil the promise, made in Section 11, to say something more about high-speed flight and the special significance of the speed of sound. It is rather a daring venture to attempt to translate the advanced treatises that have been written on this subject into the simple language that one has tried to use in this book, and one feels very humble about it, especially since one cannot claim to have had any practical experience of this kind of flight. But the duty is clear, for nowadays no book about flight—whether with or without formulae—can be complete without it. So here goes!

84. The Speed of Sound

The difference, as explained in the earlier section, is that, whereas in low-speed flight the air behaves as though it were incompressible, in high-speed flight its property of *compressibility* not only matters but becomes of extreme importance. The pressures caused by the movement of the aeroplane through the air, the pressures that are the cause of lift and drag, are communicated in all directions to the surrounding air. *The speed at which the pressure waves travel is the same as that at which sound travels in air*; this is not surprising because the travel of sound is the travel of a pressure wave. If one makes

THE SPEED OF SOUND

a noise, by clapping hands or with a pistol or whatever it may be, the air is rapidly compressed and the pressure wave travels until it strikes your ear drum thus enabling you to "hear" the noise. From the familiar examples of a gun being fired, or a bomb exploding, or thunder and lightning, we know that we see before we hear, in other words that light travels more quickly than sound. Light, at 186,000 miles per second, travels so fast that in considering examples of this kind (not of course in astronomy, etc.) the time that it takes to reach the observer can be neglected, and we can say that the difference between the time when we see a thing happen and the time when we hear it happen represents the time taken by the sound to travel from the source of sound to our ears. So it is not at all difficult to get a rough estimate for ourselves as to the rate at which the sound travels and to use this for estimating the distance of the source of sound; many people, for instance, estimate the distance of a thunderstorm by counting (not always at the correct rate) the seconds between the flash and the crash. *The speed at which sound travels under normal atmospheric conditions* has, of course, been measured accurately and is found to be *about* 1,100 *ft/sec or, as near as matters,* 760 *m.p.h.;* this means that it takes rather less than 5 seconds to travel a mile.

Now, if something that makes a noise—what better example than an aeroplane?—is travelling towards you at this speed, you will not hear it coming; for the simple reason that the noise it makes is travelling at the same speed as the aeroplane. If it travels towards you faster than the speed of sound you will hear it coming after it has gone. This isn't silly, it is true; and there is an excellent practical example in the rifle bullet. It may be small comfort to a soldier to know that if he hears a bullet coming towards him he is quite safe because it has already gone past him—but there it is. And there is another point in mentioning this example in that it shows that there is no reason whatever why bodies should not travel through the air at speeds higher than that of sound, so the idea that was so often quoted in the popular Press of a kind of "impenetrable barrier" at that speed is nonsense—as we all know now, but didn't a few years ago. Rifle bullets travel at 2 or 3 times the speed of sound, and have done so for quite a long time. So now do many other things, including a number of types of aircraft.

We can only detect these sources of noise by the effect on our ears—though if an aeroplane, or a bullet, passes very close to us we may feel pressures on other parts of our body—but the air itself, through which these sound or pressure waves travel, is very sensitive to them and adjusts itself accordingly. So it is that when an aeroplane moves through the air at speeds well below that of sound the pressure waves are able to travel ahead (as well as above and below and behind). They "warn" the air that the aeroplane is coming and that, for instance, there is a high pressure underneath the wing and a low pressure above it, so it will be easier for the air to go above. As a result of this warning much of the air that would otherwise have flowed below the wing curves upwards and flows above it. The truth of this can be shown by a most delightful experiment in a smoke tunnel; a model wing section fitted with a split flap is placed in the tunnel; when the flap is lowered streams of smoke a long way in front of the wing change direction and go over the wing instead of below it; when the flap is raised they go below it again. Anyone who has seen such an experiment needs no further convincing that the air is "warned" of what is coming (Fig. 105).

These experiments can be shown both in air and water; both behave in the same way—as if they were incompressible. Other experiments and measurements confirm that there is no appreciable change in the density of the air when it flows under these conditions.

MACH NUMBERS

If the argument has been followed so far it will be quite clear that, if the aeroplane moves at the speed of sound, *the "warning" will have no time to get ahead, the air will not be*

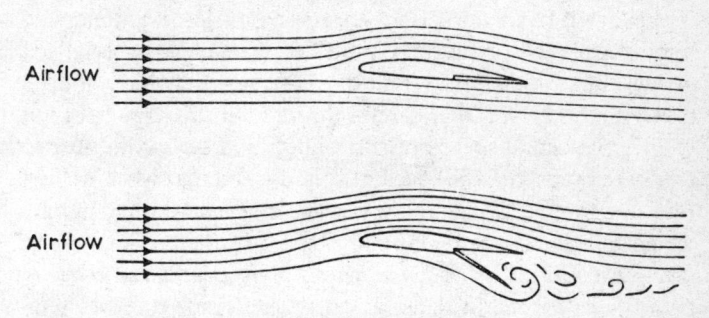

Fig. 105. Effect of lowering flap on air flow in front of aerofoil

deflected before it strikes the aeroplane, and it will come up against it with a sudden shock. That, in effect is what happens— and that is the significance of the speed of sound.

85. Mach Numbers

Since the speed of sound is so important it is sometimes convenient to speak of the speed of aeroplanes in relation to the speed of sound and to say that they are travelling at half, or three-quarters, or nine-tenths of the speed of sound, or even at the speed of sound itself or at two or three times that speed. This is expressed in terms of *Mach numbers, a Mach number of* 0·5 *simply meaning that the aeroplane is travelling at half the speed of sound.* Thus the Mach numbers in the examples given above would be, respectively, 0·5, 0·75, 0·9, 1, 2 and 3. Here, at least, is a highbrow term which anyone can understand. It is so simple, in fact, that the reader may well ask why it is necessary at all—if the speed of sound is 760 m.p.h., we know that when an aeroplane is travelling at 380 m.p.h. it is travelling at half the speed of sound; why wrap the thing in mystery by saying that it is travelling at a Mach number of 0·5?

Well, as it happens, it isn't—in this case—just an attempt to blind people with science. An observant reader—especially if he has already fallen into some of our traps—may have noticed that we have been rather careful throughout this argument not to give the actual speeds of rifle bullets and so on, but just to compare them with the speed of sound—and that when we first said that the speed of sound was, as near as matters, 760 m.p.h., we specified *under normal atmospheric conditions*. That is the clue. *The rate at which sound travels in air depends on the temperature of the air* (it depends on other things, too, but temperature is the controlling factor); *the lower the temperature the lower the speed of sound. Thus at the temperature of ground level conditions of the International Standard Atmosphere* (conditions which rarely apply in practice) *the speed of sound is about* 760 *m.p.h.; while at the temperature of the stratosphere* (*above about* 37,000 *ft*) *it is about* 660 *m.p.h.* So at 700 m.p.h. an aeroplane may be travelling below the speed of sound, at the speed of sound, or above the speed of sound, according to the temperature at the time. What matters is *not* that it is going at 700 m.p.h. but *at what fraction of the particular speed of sound it is travelling—in other words what matters is, not its speed, but its Mach number*.

When there is no need to specify the actual Mach number and we only wish to indicate that a body, or the air flow, is travelling at less than the speed of sound, at the speed of sound or above it, it is usual and convenient to use the Latin words and to speak of *subsonic, sonic, and supersonic speeds*.

As we shall soon see, it isn't just *at* the speed of sound that curious things happen, but over quite a range of speeds which include that speed, and it is useful, therefore, to introduce the word *transonic*. Our subject then falls into three quite distinct parts, i.e. flight at subsonic speeds which is what we have so far considered, flight at transonic speeds which has problems all of its own, and flight at supersonic speeds in which we are in a new world altogether and all the rules are so much the opposite from what we have already learnt that it reminds us of *Alice Through the Looking-glass*.

86. Flight at Transonic Speeds

But Alice was fortunate in that in her dream the barrier, in her case the looking-glass, melted away, and she found herself all of a sudden in the new world. It was not quite so easy to get through the barrier which divides flight at subsonic speeds from flight at supersonic speeds. Not unnaturally we approached this barrier with considerable caution. We didn't quite know what was going to happen, and what did happen was apt to be alarming—all the more so because it differed in different types of aircraft.

Perhaps we didn't know it at the time, but it was the near side of the barrier, the approach to the speed of sound, that proved the most difficult to negotiate.

We were also very much in the dark because, just when the wind tunnel might have been most helpful, it came up against the barrier itself—it got choked, and we couldn't reach the speed of sound at all. It was more difficult to get through the speed of sound in wind tunnels than in actual flight.

Even now we know less about flight at transonic speeds than we do about flight at supersonic speeds, and of course we cannot fly at supersonic speeds without first going through the barrier, though if there is plenty of power in hand there is something to be said for getting through it quickly. But first let us take it steadily, for we shall then see more clearly the significance of the range of speeds which we class as transonic.

87. Shock Waves

Now, what happens when we approach the speed of sound? All we have said so far is that the air is no longer warned of our approach and so it comes up against the aeroplane with a shock. The evidence of this is the appearance of a *shock wave, a sharp dividing line going out from the surface of the body or wing and representing a sudden drop in the speed, and an increase in the pressure and density of the air* (Fig. 106). These

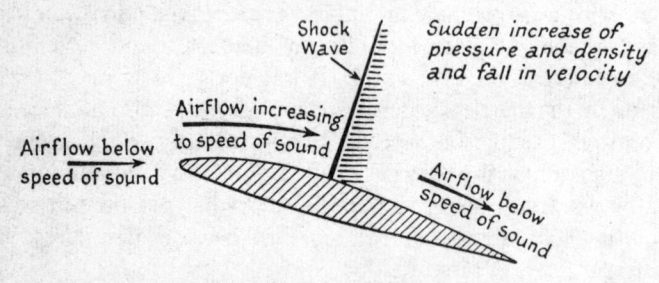

Fig. 106. A shock wave

waves are not usually visible to the naked eye, but they can be photographed by certain optical means. But, visible or not, they are very real and very important. A blast wave from an explosion is a very intense shock wave, and this, of course, can do actual damage to obstacles in its path.

88. The Shock Stall

What is the practical significance of shock waves? What effect do they have on the performance or behaviour of the aeroplane?

Such questions are soon answered. The wave is itself a barrier to the smooth flow of air over the surface; thinking back to the analogy of Section 17, the shock wave is an adverse

THE SHOCK STALL

pressure gradient of a most vicious kind, a hill so steep that it is almost vertical. But, whereas with the speeds we were considering in Section 17 the air was warned that there was a hill ahead, and if the hill was too steep it wouldn't even approach the hill, at the speeds we are now considering it gets no such

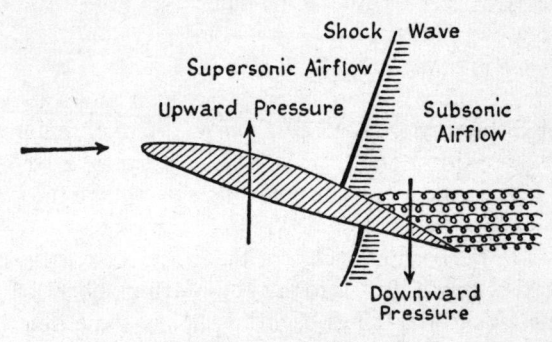

Fig. 107. **The shock stall**

Note: 1. The turbulent flow behind the shock wave on the upper surface.
2. The new rather feeble shock wave starting to form on the lower surface.
3. The downward pressure at the rear of the wing caused by the increased pressure behind the shock wave on the upper surface and resulting in a decrease in lift.

warning, it even increases its speed downhill—and then comes the shock.

No wonder that behind the shock wave the air near the surface of the wing becomes violently turbulent and the whole distribution of pressure over the wing is changed, resulting in a loss of lift and increase in drag (Fig. 107). Very like a stall—yes, so like one that the result is called a *shock stall*. And the similarity does not end at what happens to the air flow, the pilot too experiences much the same effects on the aeroplane as those with which he is so familiar at the ordinary stall, buffeting and shaking of the airframe, change of trim (the nose usually dropping but sometimes tending to rise) and loss of control. There are differences, of course. Whereas in the ordinary stall the controls become sloppy and ineffective, in the shock stall they may become so stiff as to be impossible to move, and if a device, such as a tab, is used to move them, the effect may be so violent as to be dangerous. The main difference, however, is that *whereas the ordinary stall occurs at a large angle of attack, the shock stall may occur at any angle of attack* and is more likely to occur at the small angles associated with high speed. So, to recover from an ordinary stall we must put the nose down and increase speed, but to recover from a shock stall we must reduce speed either by pulling the nose up or, better still, by using some kind of air brake.

It has so far been implied that the shock wave is formed at the speed of sound, that is to say at a Mach number of 1. But, as a matter of fact, things begin to happen some time before this—and for a very good reason. As was explained in some of the earlier parts of the book, we get a decreased pressure in a venturi tube or on the upper surface of a wing section because the air flow is speeded up by the shaping of the surfaces, and thus it is that, even before the aeroplane itself is travelling at the speed of sound, the air flow over some part of it will reach that speed and cause a local shock wave. This may occur in many places, such as on the top of a wing section, or on some bulge on the body or elsewhere, or at the entry to a jet engine.

The Mach number at which this occurs is called the *critical Mach number of the aeroplane*. The more cambered the surface, the more bulbous the body and general shape of the aeroplane, the more likely is it that the speed somewhere will exceed the speed of sound and so the lower will be the critical Mach number. It is clearly an advantage to have a high critical Mach number, and so to be able to go as near as possible to the speed of sound without experiencing a local shock stall, and it is the long slim aircraft that displays this characteristic. The good old Spitfire provided an interesting example of a high critical Mach number (nearly 0·9 in some of its versions), interesting because at the time of its original design such speeds were hardly contemplated.

Whatever the difference or similarities between the ordinary stall and the shock stall, it was customary at one time to think of them as the limits to flight in either direction—if we tried to fly too slowly, we stalled; if we tried to fly too fast, we stalled. The margin between them in normal ground level conditions is likely to be considerable, say 100 m.p.h. for the ordinary stall and 600 m.p.h. for the shock stall (this assumes a critical Mach number of 0·8). But at 40,000 ft the ordinary stalling speed of 100 m.p.h. has become 200 m.p.h. (owing to the fall in density), and in a tight turn at $4g$ this will become 400 m.p.h. (owing to the increase in loading). At the same time the shock stalling speed of 600 m.p.h. has become nearer 500 m.p.h. (owing to the drop in temperature) and even less because in a tight turn the shock stalling speed is likely to be reduced. Thus in these by no means impossible conditions the range of speeds at which we can fly, assuming that we cannot go beyond the stall in either direction, has become very small and we are in a dilemma between flying too slowly and flying too fast—perhaps the devil and the deep sea is an apt metaphor! At even greater heights the margin becomes even further reduced until we reach a new kind of ceiling; the other ceiling (of Section 69) could be raised by increase of power, this one can only be raised if we can penetrate the barrier of the speed of sound.

89. Wave Drag

For we can no longer make the assumption that the shock stall, or the speed of sound, is a barrier beyond which we cannot go. Indeed we ought never to have made such an assumption with the example of rifle bullets as evidence that this barrier is not impenetrable.

In the last section it was stated that the critical Mach number of an aeroplane was that at which the shock stall occurred; it was also stated that when the shock stall occurred there was a loss of lift and an increase of drag. Like so many definitions in this subject there is a certain amount of vagueness about the definition of critical Mach number, and people are inclined to interpret it according to their own point of view. If the aeroplane suddenly starts to judder violently, or to drop its nose, or if the controls suddenly seem to be locked, the pilot rather naturally takes this as indicating that he has reached the critical Mach number; but to the theoretical man, plotting his graphs on a sheet of paper, things seem rather different— he won't feel any judder, his nose won't drop or his controls become locked, all he will see is some sudden kink in the curve, some sudden rise in drag or drop in lift; and it is quite likely that this will not occur at exactly the same Mach number as that at which the pilot feels things happening.

But in a book of this kind we are not concerned with exact definitions; we are much more concerned with what happens and why. So let us look a little more closely at the effect of the formation of the shock wave. When the airstream hits the shock wave it loses speed and changes direction, and as already stated there is a sudden increase in pressure and density —but there is also an increase in temperature. This means loss of energy, and we have to keep adding energy to make up the loss; in other words we need extra thrust to counteract the extra drag. The shock wave, in short, is the cause of a new kind of drag called *wave drag;* it first appears at the shock stall or critical Mach number, it increases rapidly as the shock waves grow stronger, and it is a major obstacle not only in getting through the speed of sound but in the maintenance of supersonic flight once we have got through.

WAVE DRAG

But this is not all. The sudden rise of pressure in the shock wave upsets the flow of air in the boundary layer, causing it to separate from the surface and become turbulent over the remainder of the surface (Fig. 107). This is yet another source of drag—called *boundary layer drag* or *separation drag*—which, though quite different in its nature from wave drag, is

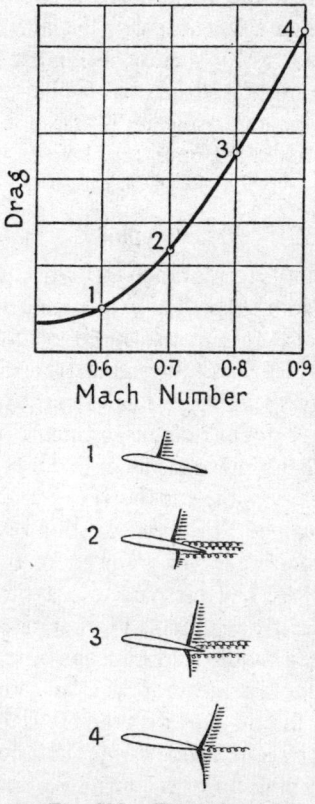

Fig. 108. **Shock drag**

Note how the drag increases as the shock-wave pattern changes.

a direct result of the formation of the shock wave. The two added together, wave drag plus boundary layer drag, form what is sometimes called *shock drag*—and the increase in total drag resulting from the addition of this new source of drag as we approach the sound barrier is enormous, and may represent a ten times increase (Fig. 108).

The loss of lift resulting from the formation of a shock wave on the top surface of a wing is easily explained by the increase of pressure, and so loss of suction, behind the wave; and the juddering and changes of trim, by oscillations of the shock wave to and fro on the wing and corresponding changes in the extent of the turbulent boundary layer.

90. Sweepback

In our earlier attempts to approach the barrier we naturally concentrated on avoiding the shock stall—and its consequences. That is why for attempts at the speed record we waited for hot days; on the face of it rather an extraordinary thing to do because, generally speaking, we can expect to get better performance out of both engine and aeroplane in cold dense air. But success at that time depended upon keeping the Mach number as low as possible for the speed attained, and as has already been explained, the speed of sound is higher in warm air, and so the Mach number is lower for the same airspeed.

The designer, too, had his part to play in the approach to the barrier. He already knew the first need—*slimness*—because this had been required as a means of reducing drag even with the old-fashioned ideas of what constituted high-speed flight, 300 or 400 m.p.h. But with the realization of the significance of shock waves and shock drag, and how they first arise owing to the speeding up of the airflow over a cambered surface, it becomes more than ever necessary in approaching the barrier *to keep the camber low, to have wings with a low ratio of*

SWEEPBACK

thickness to chord, to have laminar flow aerofoils, slim fuselages, and smooth surfaces free of bumps and bulges.

But the designer's most significant contribution to the achievement of transonic flight was the introduction of *sweepback*, the sweeping back of the wings, not by just a few degrees as had been sometimes done in the past for reasons of stability, or for improving the pilot's view, but by 20°, 30°, 45°, or even more (Fig. 109 and Plates 22 and 23). By this means in one

Fig. 109. **Sweepback for transonic speeds**

fell swoop, as it were—or should we say sweep—he not only delayed the shock stall and made the approach to it more gentle, but reduced its severity when it did come. Generally speaking, too, the greater the sweepback the greater is the effect. How and why does this happen? Well, as so often, a simple explanation may be criticized as not being entirely correct, but in this case it isn't far wrong and accords fairly well with the facts.

The speed of air flowing over a swept-back wing can be considered as having two parts, or components, i.e. that which flows across the chord of the wing (at right angles to the leading edge), and that which flows along the span of the wing towards the wing-tip (parallel to the leading edge). Fig. 110 shows these two parts, and how the greater the sweepback the less is the airspeed across the chord of the wing compared with the speed of the aircraft as a whole. Now the formation of

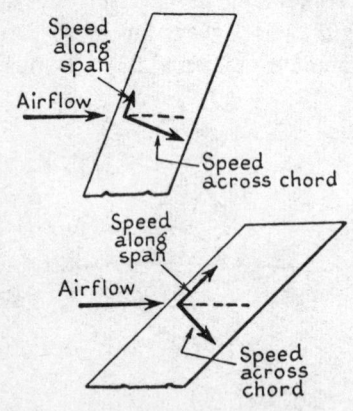

Fig. 110. **How sweepback reduces shock drag**
Note how the speed of flow across the chord
is reduced with greater sweepback.

the shock wave, and its severity, depend on the Mach number of the airflow across the wing, and so, *other things being equal* (our old friend again), *the greater the sweepback the nearer we can approach to a Mach number of one without encountering a shock stall.*

91. Vortex Generators

Brief mention must be made of small projections on certain parts of high-speed aircraft which some observant readers may have noticed—and may well have puzzled over. These small projections, called *vortex generators*, are of various shapes but

VORTEX GENERATORS

generally either of the wedge or ramp type, or of the bent-tin type (Fig. 111); and they are situated at rather surprising places such as on the top surface of a wing not far behind the leading edge. Their name, "vortex generators," aptly describes their function—but surely the last thing we want to do is to generate vortices, especially when we want to fly fast! It is rather strange, and the explanation really lies in the small

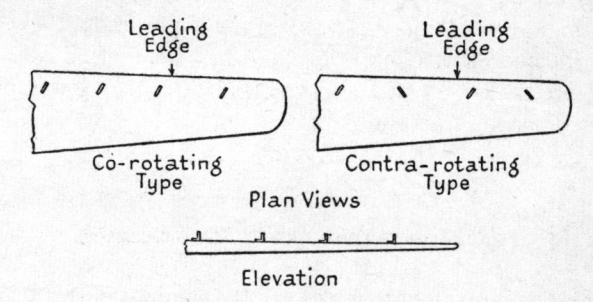

Fig. 111. **Vortex generators**

height to which these devices project, which is only two or three times the thickness of the boundary layer. What they do in effect is to put new life into a sluggish boundary layer, stir it up and give it extra energy by inducing some from the faster surrounding air, and so enable it to get farther along the surface before it is slowed up and separates—in short, they have very much the same effect as other means of controlling the boundary layer, e.g. by blowing or sucking, and of course they are much simpler.

Other methods of weakening the shock waves, and of reducing the tendency of the air in the boundary layer to separate from the surface, include thickening of the trailing edge of the wing, and fitting at the trailing edge streamline bodies, called rather disrespectfully "carrots," but more correctly *Küchemann bumps*.

92. Wing and Body Shapes

The need for thin wings and slim bodies for flight at transonic speeds has already been emphasized; but there are also other considerations affecting the shape of wings and bodies. Leading edges of wings, and noses of fuselages, can be much sharper than we have become accustomed to in the old streamline shape; strange as it may seem the high-speed airflow can turn the corner at a sharp leading edge, even when the wing is at an angle of attack, more easily than the slower flow. A good shape of aerofoil section is one of bi-convex camber,

Fig. 112. **A wing section for transonic speeds**

with little if any difference between the cambers of the top and bottom surfaces, and symmetrical in the fore and aft directions, i.e. with the front half the same shape as the rear half (Fig. 112).

Sweepback will, of course, affect the plan form of the wings, but it is the sweepback of the leading edge which is most important; the trailing edge may be parallel to the leading edge, but for structural and other reasons usually has less sweepback than the leading edge (Plates 17 and 25), giving a tapering plan form, and in the extreme may even be at right angles to the body, giving the so-called *delta* shape (Fig. 113). This, with its aspect ratio of 1 or less, is a sharp reminder that in attacking the sound barrier we have forgotten all about induced drag, economical flying—and all that sort of thing. Sweepback, and even the delta shape, can be applied to the tail, and to the vertical surfaces such as fin and rudder, as well as to the wings; but of course the delta shape in itself can take the form of a flying wing, and there is then no need for a separate tail unit (Plates 20 and 23).

Attempts to reduce shock drag have resulted in the bodies of some high-speed aircraft having acquired a waist line in accordance with the *area rule* (Plate 61). This is a term which has now come into common use—and misuse—so we should at least give it a mention. Like many principles of design it is rather more complicated, and has more applications, than is commonly supposed, but fundamentally it is a means of avoiding the bulges which cause shocks by keeping the cross-sectional area of the aircraft as nearly constant as possible,

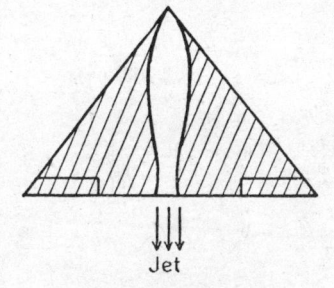

Fig. 113. **Delta wing for supersonic flight**

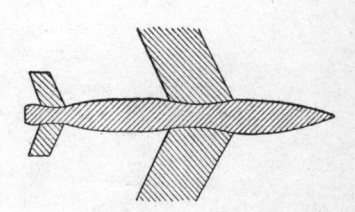

Fig. 114. **Showing the effect of area rule on a fuselage**

or at least changing only smoothly and gradually, as we go from nose to tail. So when we get to the wings the cross-sectional area of the fuselage is reduced, hence the waist, and expanded again after the wings (Fig. 114). If the wings are swept back the waisting can be more gradual. The same applies at the tail or other necessary protuberances.

93. Through the Barrier—and Beyond

The term "barrier" is quite a good one, perhaps "hurdle" would be even better, for the greatest difficulties lie in getting through it, or getting over it—according to which metaphor

FLIGHT WITHOUT FORMULAE

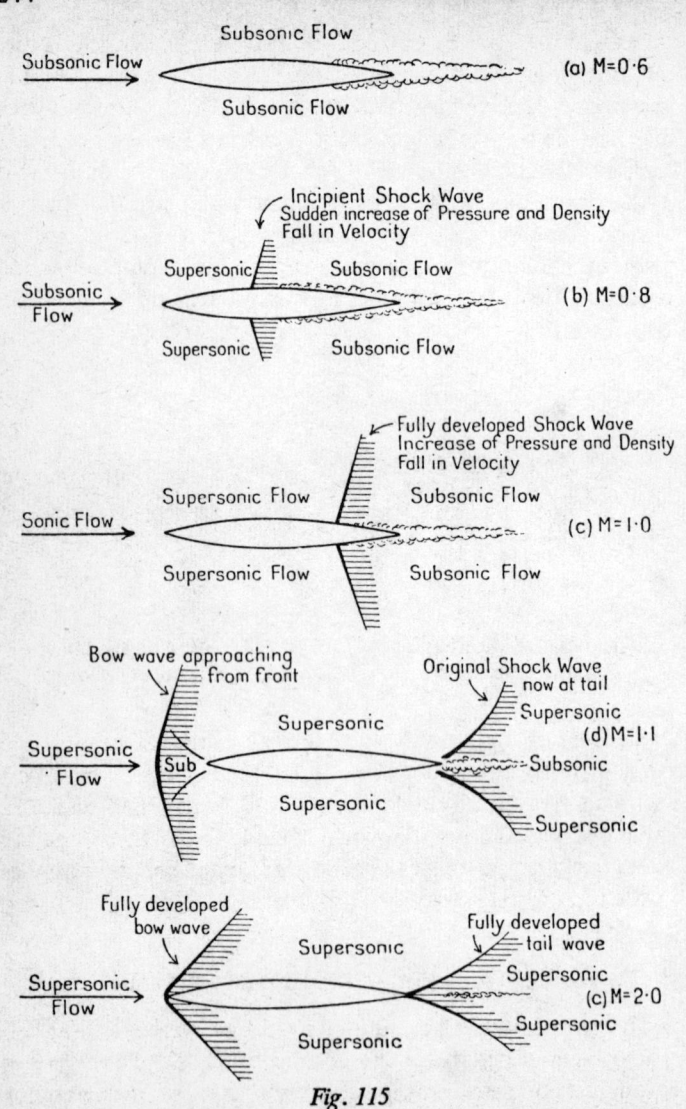

Subsonic Flow →
Subsonic Flow
Subsonic Flow
(a) M = 0.6

Incipient Shock Wave
Sudden increase of Pressure and Density
Fall in Velocity

Subsonic Flow →
Supersonic | Subsonic Flow
Supersonic | Subsonic Flow
(b) M = 0.8

Fully developed Shock Wave
Increase of Pressure and Density
Fall in Velocity

Sonic Flow →
Supersonic Flow | Subsonic Flow
Supersonic Flow | Subsonic Flow
(c) M = 1.0

Bow wave approaching from front
Original Shock Wave now at tail

Supersonic Flow →
Sub — Supersonic — Supersonic
Supersonic — Subsonic
Supersonic
(d) M = 1.1

Fully developed bow wave
Fully developed tail wave

Supersonic Flow →
Supersonic
Supersonic
Supersonic
Supersonic
(e) M = 2.0

Fig. 115

we prefer. It is at the barrier that we meet the shock wave for the first time, the changes in trim, the juddering, the sudden rise in drag and loss in lift. On the far side we find ourselves in a strange new world, a world in which we are all beginners and have to learn about flight, both in theory and practice, all over again—the land of *compressibility*.

And, just as a whole book was needed to describe what Alice found in looking-glass land, so a whole book would be needed to describe this land of compressibility. Several such books have now been written, but perhaps in these last few sections of our book we can at least get some idea of the fascination of the study of this new kind of flight.

During the passage through the barrier, i.e. at transonic speeds, or between Mach numbers of about 0·75 and 1·2, there have been sudden changes of drag and lift. These have been caused by the formation and movement of shock waves, together with the effects of the shock wave on flow in the boundary layer. Fig. 115 shows in pictorial form typical changes in shock wave pattern as we pass through the barrier. A small shock wave first appears on the upper surface of the wing at about the highest point of the camber—this is what we would expect since this is where the air is flowing fastest. As the Mach number increases, this shock wave becomes more intense, extends farther from the aerofoil surface and moves backwards. Then another shock wave starts to form on the lower surface—rather farther back than the top one, again as we would expect. With still further increase of Mach number both the shock waves move farther back until they reach the trailing edge. Then, soon after the whole wing is moving above the speed of sound, i.e. at a Mach number greater than 1, another shock wave appears ahead of the leading edge. After this, further increases of speed have little effect on the shock wave pattern except that the shock waves take up a more acute angle to the surface of the wing, and the bow wave may become attached to the leading edge. These are indications that the whole flow is supersonic, both in front of the waves and behind them, whereas in transonic flight the flow behind the shock wave is subsonic.

Figs. 116 and 117 show the changes of lift and drag as we pass through the barrier; these are caused by the changes

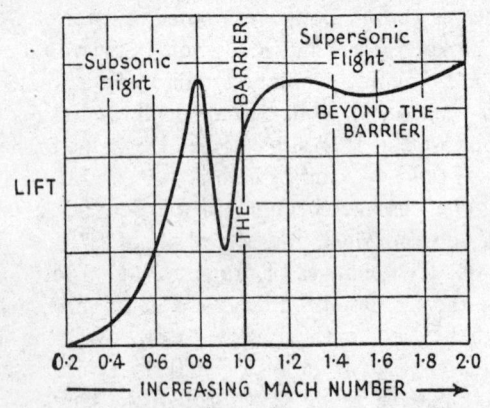

Fig. 116. How the lift changes through the barrier

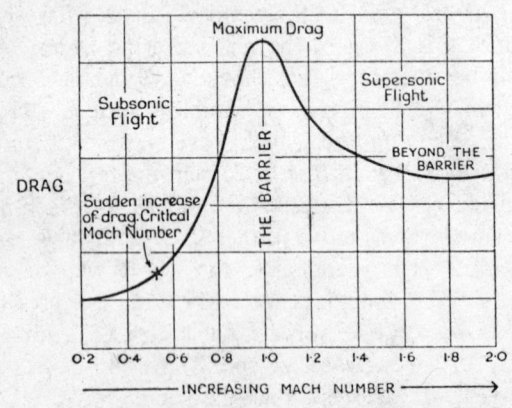

Fig. 117. How the drag changes through the barrier

SUPERSONIC FLOW

in shock wave pattern, which in turn cause pressure changes over the wing, and these in their turn are what affect the behaviour of the aeroplane.

The comparative stability of the shock pattern on the far side of the barrier is likewise reflected in the curves of lift and drag which no longer fluctuate violently, and in the behaviour of the aeroplane which no longer suffers juddering or violent changes of trim. In short, flight at supersonic speeds is steadier than flight at transonic speeds; we are right in it, as it were, not half in, and provided we learn our new subject and design aeroplanes to suit the new conditions there is no particular difficulty and we know what to expect.

94. Supersonic Flow

But we are in a new world and we certainly have a new subject to learn. *Air flowing at supersonic speed doesn't know what is coming; the flow is unaffected until it reaches an object*—how different from flow at subsonic speeds which changed direction some distance in front of the object! *Supersonic flow likes to meet sharp-pointed obstacles*, and *doesn't object to turning sharp corners*—what a contrast to the streamline shape and smoothly rounded corners favoured by subsonic flow! But perhaps the most striking contrast is in the flow through a shape such as a venturi tube in which we have a gradually contracting duct followed by an expanding duct. Unlike the subsonic flow, which likes to enter the venturi where it speeds up as it runs "downhill" as its pressure drops, *supersonic air flow strongly objects to the converging duct, shock waves are formed and the flow may be choked altogether*—this was the problem in the early attempts to design supersonic wind tunnels—and if the air does go through *it slows down and its pressure rises*. On the far side of the venturi, in the expanding duct in which the subsonic flow slowed down and the pressure increased as it laboured "uphill," *supersonic flow speeds up and its pressure decreases*. We have probably by now become so accustomed to the venturi tube idea for subsonic flow that we accept it as natural and common sense, but when we first described it we remarked that some people thought it was contrary to common sense, and the reader may have thought so himself; but it is interesting to note that supersonic flow behaves more in accordance with what one might expect from common-sense ideas—and the same, of course, applies to sharp-pointed obstacles as against streamline shapes.

Notice, however, that *in both supersonic and subsonic flow increasing speed goes with decreasing pressure*—good old Bernoulli!

95. Supersonic Shapes

The way in which air flows through contracting and expanding ducts at supersonic speeds accounts for the internal shapes of ram-jets (Fig. 57) and gas turbines (Fig. 58); shapes which,

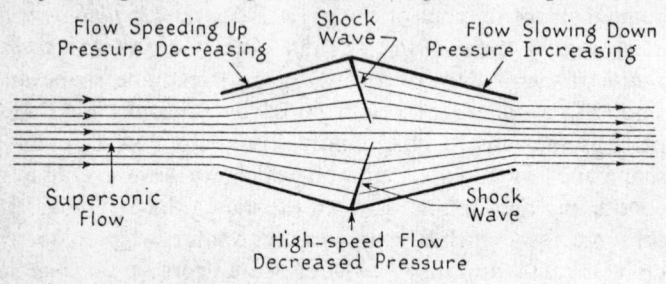

Fig. 118. **A supersonic venturi tube**

except for slight variations, expand and then contract—just the opposite to a venturi. In effect a kind of inverse venturi tube (Fig. 118) is needed to give the same results in supersonic

SUPERSONIC SHAPES 249

flow as a venturi tube gives in subsonic flow. The new kind of air flow also accounts for the cross-sectional shape of aerofoils designed for supersonic speeds. Some of these are illustrated in Fig. 119, and although there is not much to choose

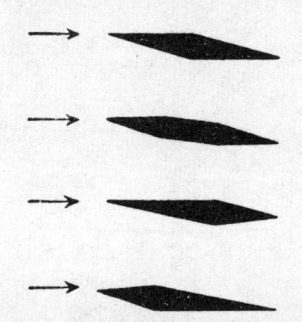

Fig. 119. **Supersonic aerofoil shapes**

between them the bottom one is probably the best of all—it would be difficult to imagine shapes less suitable for subsonic flow. Fig. 120 shows how air at supersonic speeds flows over a double-wedge supersonic aerofoil, which is here set at an angle to give the best ratio of lift to drag.

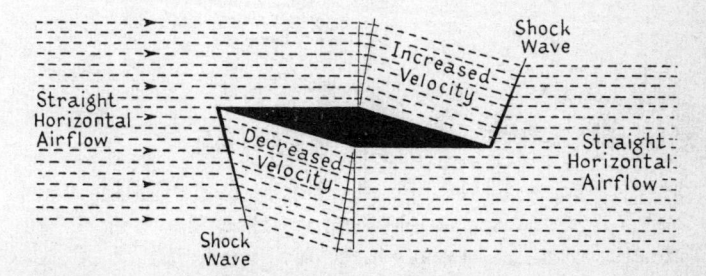

Fig. 120. **Supersonic airflow over a supersonic aerofoil inclined at a small angle of attack**
Contrast with Fig. 24, showing subsonic flow over a subsonic aerofoil.

But whereas in subsonic flow the cross-sectional shape is more important than the plan form, in supersonic flow the plan form is more important than the cross-section. At low supersonic speeds, if that is not a contradiction in terms, the

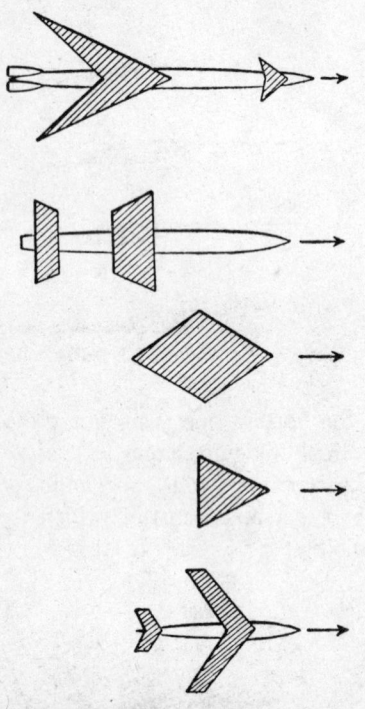

Fig. 121. **Supersonic plan shapes**

heavily swept-back leading edge retains the advantage that it had at transonic speeds and the plan shapes shown in Fig. 121 are all possibles. At really high Mach numbers such as occur with missiles the rectangular wing may come into its own again, and there are advantages in square-shaped wing tips. Poor old aspect ratio has been forgotten, and with it all ideas of economical flying; it might almost seem that lift has been forgotten, and certainly most of the weight lifting is now done by thrust rather than by the lift of the wings.

96. Sonic Bangs

So far, we have only considered the effect of shock waves on the flight of the aeroplane—but that is by no means the whole story. It will be noticed in the diagrams that the shock waves, the sudden increases of pressure, although they occur and may be most violent at the surface of the wing or other parts, extend outwards from the surface, becoming less severe, it is true, but none the less having an effect at considerable distances from the aircraft.

If an aeroplane flying straight and level at supersonic speed were to fly close over the head of an observer on the ground—which God forbid!—the observer would be hit, more or less literally in this case, first by the bow shock wave, then by the trailing-edge shock wave, then by one or two shock waves from the tail plane, all in quick succession, and interspersed by a few more from the body and other parts. Two or three of these bangs might stand out from the rest, but more probably it would all sound like a roll of thunder, and damage might be done by the shock, not only to people but even to buildings. These are the *sonic bangs* that we hear so much about, even if we haven't experienced them ourselves, and they pose one of the most serious problems in connection with supersonic flight.

There is nothing very new about sonic bangs: the crack of a rifle bullet, the thunder caused by a lightning flash, the the blast of an explosion and even the crack of a whip are all examples of the same phenomenon; but these are only isolated incidents, unlikely to occur at frequent intervals, and not usually of such intensity as to break windows, shatter greenhouses or even disturb one's sleep.

The bangs caused by supersonic aircraft are, however, a very different matter, and seem likely to become such a nuisance that they may hinder progress in the development of such aircraft, more especially the supersonic airliner like the Anglo-French Concorde (Plate 32). Of course, in considering the effect of an aircraft flying low and directly overhead, we have taken an extreme case to illustrate the point; and, as one would expect, the bangs are less severe, more dispersed and less distinct the farther one is from the aircraft, in other words, the higher it is flying and the greater the distance its flight path from being directly overhead. But even from considerable heights the bangs can be objectionable, as has been amply proved by experiments over selected areas both in this country and in the United States. Unfortunately too the greater the speed the greater are the bangs and, what is not so obvious perhaps, the greater the weight of the aircraft—or what comes to the same thing, the more it manoeuvres and increases the wing loading—the more severe is the effect.

Can anything be done about it? Unfortunately the answer is, Not much—short of not flying at supersonic speed! For the immediate future the only course seems to be to limit speeds until great heights are reached, or perhaps even to prohibit supersonic flight over land at all. It seems rather a drastic remedy, but there does not appear to be any alternative unless the public can be persuaded to put up with the nuisance in the cause of progress (?)—after all, we have already had to put up with a good deal in this cause.

97. Other Problems of Supersonic Flight

Though the conditions of supersonic flight are more steady and more predictable than those of transonic flight, the drag is of course very high, which means that great thrust is required to maintain flight. The propeller is not a practical proposition at these speeds, and we have to rely on the jet engine, if not the ram-jet and the rocket—eventually perhaps atomic energy.

The controls are too heavy for manual operation, and power-driven controls become a virtual necessity—usually with some artificial "feel" introduced to make the pilot feel at home.

Another difficulty arises because the control surfaces themselves are not so effective as in subsonic flight—this for two reasons: first, that they are working in that part of the air flow which tends to separate from the surface; and secondly, because they do not have any effect on the flow over the surface in front of them—this flow doesn't even know when the control surfaces are moved—and it will be remembered that the main effect of moving the control surfaces at subsonic speeds is to alter the flow, and the pressures, over the *front* parts of the surface.

At Mach numbers of about 2 an entirely new problem arises, due to the rise in temperature caused by the motion of the aeroplane through the air and the consequent skin friction; this essentially practical problem, sometimes rather misleadingly called the *heat barrier*, may prove very difficult to solve, and may be the limiting factor in the speeds which can be achieved in the next few years, even if we learn to accept the sonic bangs. No doubt its impact will be softened by insulation of the surfaces, by improved materials used for the surface skin, and by refrigeration devices; but the term "heat barrier" is misleading because, unlike the sound barrier, it is not just a hurdle, something to be got over and left behind; it gets worse and worse as we go faster and faster—it will always be with us when we fly at high speeds in the atmosphere. Fortunately, however, it will not, like sonic bangs, affect people on the ground.

The frailty of the human body may be another limiting factor, though more from the point of view of the accelerations involved in reaching these speeds, or in slowing down again, or in manoeuvres; the human body doesn't seem to provide any serious problem in regard to steady flight—at any rate at any "reasonable" speeds.

But perhaps the greatest problem of supersonic flight, and the one that we are most likely to forget, is that aircraft cannot reach this condition of flight without going through all the stages of subsonic and transonic flight—and we know enough

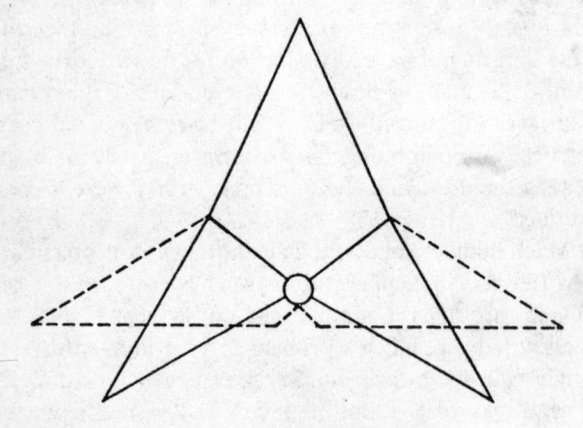

Fig. 122. **Variable sweep**

now to realize that an aeroplane designed for supersonic flight could hardly be more unsuitable for subsonic flight, particularly for take-off and landing.

Unlike sonic bangs and heat barriers, this problem is not insuperable—in fact, to some extent it has already been solved by the use of what has been termed *variable geometry*, the use of wings which can be swung on hinges from a position of little or no sweepback (and reasonably high aspect ratio) for low-speed flight, to right back to form a delta wing with much less resistance and small aspect ratio for high-speed flight (Fig. 122 and Plate 63). Other forms of variable geometry may also be tried, extension of the wings either in span or chord, and variable camber or new types of flap.

The rocket or missile gets over this difficulty by its vertical or near-vertical take-off, by passing so quickly through the subsonic and transonic stages, and by our lack of concern whether it returns to earth whole or in bits, or even at all! Maybe the missile points the way for the manned aircraft of the future, but those who man it will want these problems solved first.

98. The Future

So to the future. There is little to add to what has already been said. There are certainly problems to be faced, problems to be solved; there always have been and, thank God, there always will be. We have come a long way—in 1908 the world speed record was 50 m.p.h., in 1967 it was over 4,500 m.p.h. (or, if we include spacecraft, over 24,000 m.p.h.)—and this increase of speed is only a reflection of all the many problems which have been faced and solved. In this book I have brought you, as it were, along this road, I have shown you the problems on the way, and I have tried to explain how they have been solved; we have surmounted the first serious obstacle, the sound barrier, and we are now tackling another, perhaps more serious, the heat barrier.

What speeds will be attained in the next few years? What will be record speeds when the next edition of this book is published? In the last edition we said that we had already got as far as giving a new name to speeds above a Mach number of 5, but it is now more than a name, it is an accomplished fact, for the latest speed records are in the region of *hypersonic* speeds. Photographs of shock waves have been obtained at Mach numbers of 10 and more, and except that the waves lie back at very acute angles, the pattern is much the same

(Fig. 123). Rocket-driven spacecraft have, of course, far exceeded even these speeds, and do not seem to have encountered

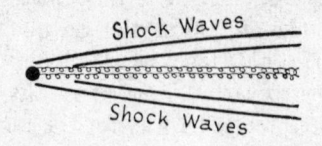

Fig. 123. **Shock waves at hypersonic speeds**
Sketch made from a photograph of a sphere moving at a Mach number of 10.

any new problem so far as what may be called the principles of flight are concerned, but we shall have a little to say about these in the next section.

99. Into Space

This book, so far, has been mainly concerned with the flight of what we defined at the outset as "aeroplanes," but what about *missiles*, *satellites* and *spaceships* (Plates 45 and 47)? They are not aeroplanes, not even aircraft (as we have defined the term), and although one may even question whether they fly, in the strict sense of the word, no book on flight, with or without formulae, can any longer leave them out of consideration altogether.

So let us consider first what happens to the most familiar and simple of missiles, such as a cricket ball thrown into the air. If it is thrown vertically upwards it will gradually lose its upward velocity (due to the force of gravity), stop, and then come down with increasing velocity to where it started from. The height it reaches and the time of flight will depend upon the velocity with which it was projected upwards.

Now what happens if the cricket ball is thrown horizontally? It will immediately begin to acquire a downward velocity,

INTO SPACE

and to fall with increasing downwards velocity, while its horizontal velocity—neglecting the gradual slowing up owing to air resistance—will remain constant. Its path of flight will look something like those shown in Fig. 124.

It will be quite clear that the distance the cricket ball will travel over the ground before striking the ground will depend

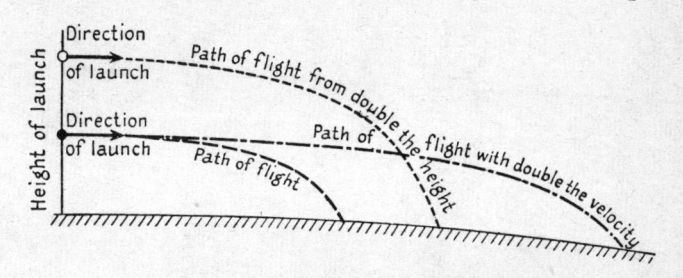

Fig. 124. **Bodies launched horizontally**

on the height at which it is projected, and the velocity with which it is projected.

It will be quite clear, too, that, no matter what the height or what the velocity with which it is projected, *if the earth were flat* it would strike the ground sooner or later. *But the earth is not flat—and this makes all the difference, because as the body falls the earth is curving away from it.* Of course the practical difference is negligible so long as we think of cricket balls; it is in fact negligible even at the speeds of aeroplanes, and has little effect on the range of shells and rifle bullets with their muzzle velocities of the order of 1,000 or 2,000 m.p.h. But the development of rockets has made possible the achievement of speeds of several thousand miles per hour; and, more important, it has brought within reach heights far beyond those achieved by aeroplanes, and where there is little or no air resistance, and so it is much easier both to attain and to maintain such speeds.

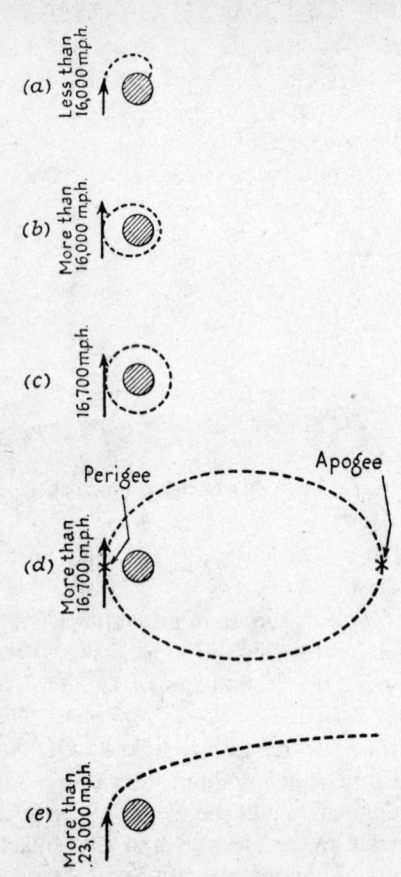

Fig. 125 **Satellites**

Speeds refer to horizontal launches from 500 miles above the earth's surface. In the interests of clarity this distance has been exaggerated in comparison with the radius of the earth (about 4,000 miles).

INTO SPACE 259

So let us think now of a missile being launched horizontally at a height of say 500 miles above the earth's surface. This height is chosen as being well clear of the earth's atmosphere so that the effects of air resistance are negligible. The missile must first be propelled to this height by a rocket, or rather a series of rockets, and if then it is finally launched, "let go" as it were, so that there is no more rocket power behind it, it will behave just like the cricket ball in that it will gradually acquire a downward velocity while its horizontal velocity remains constant—much more constant than the cricket ball owing to the lack of air resistance. But the scale is so different, it is now 500 miles up, and the horizontal speed may be 5,000, 10,000 or more m.p.h., and as it falls towards the earth, the earth is curving away appreciably so increasing the range it will travel over the earth's surface (Fig. 125a). At a certain speed—about 16,000 m.p.h. when launched from this height—it will not come really close to the ground until it is half way round the earth, and it will by then have acquired extra speed (or, to be more correct, it would have acquired extra speed if it hadn't encountered the earth's atmosphere), and will continue to clear the earth and travel round the other side to its launching point again—falling all the time towards the earth, but on this part of its path actually getting farther away from it all the time! (Fig. 125b). It is an artificial satellite—but not a practical one, because as it encountered the earth's atmosphere on the far side it would lose its energy and really fall to the earth.

But if the launching speed is increased above 16,000 m.p.h. the missile will more easily clear the far side until at a speed of about 16,700 m.p.h. it will travel on a circular path round the earth, keeping about 500 miles from it all the way round, and so well clear of the atmosphere (Fig. 125c); it is, as it were, on the end of a string (500 miles long), the pull in the string (the force of gravity) providing the centripetal force, which is

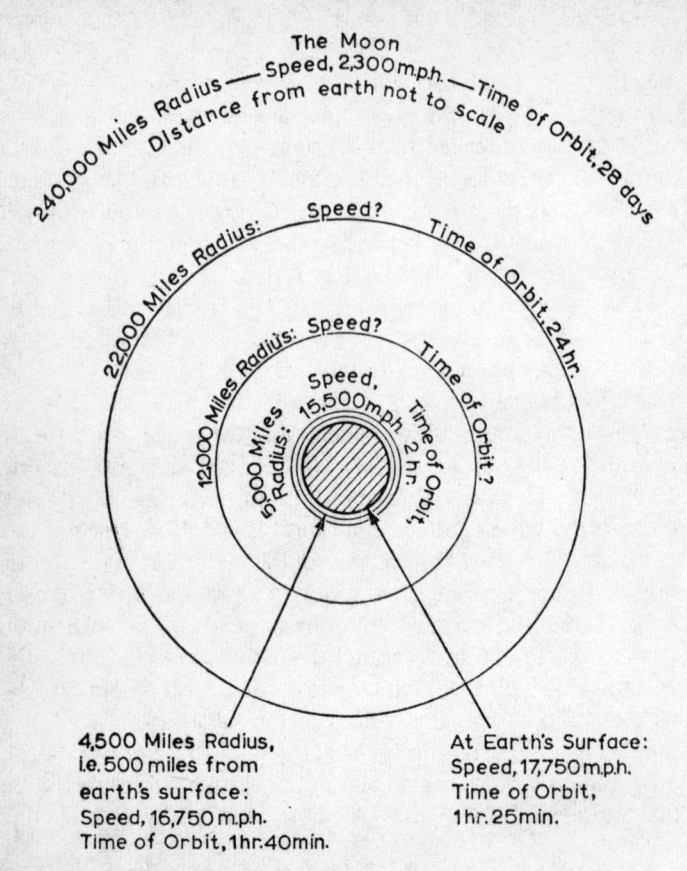

Fig. 126. **Circular orbits at different distances from the centre of the earth**

Note: It is left to the reader to fill in the speeds, and time of orbit, where queried.

balanced by the centrifugal force, so that in effect it is "weightless". It is now a practical satellite, and the velocity of launch to achieve this state of affairs is called the *circular velocity*.

Most of the spacecraft and satellites so far launched have been sent off at speeds rather above the circular velocity—to make quite sure that they get round the far side. Fig. 125*d* shows the effect of this; as the launching speed is increased the path becomes a more and more elongated ellipse, and the farthest point (called the *apogee*) may be thousands of miles away from the earth. The point of nearest approach to the earth is called the *perigee*.

At a certain speed, about 23,600 m.p.h. from 500 miles, the ellipse doesn't close at all, it becomes an open curve and the missile "escapes" from the earth and its gravity (Fig. 125*e*)—it will travel through space until it comes under the influence of some other heavenly body such as the sun. The velocity which results in this is called the *escape velocity*.

Now, these critical velocities, circular velocity, escape velocity, and so on, decrease as the distance from the earth's surface of the launch increases (Fig. 126). Thus the circular velocity near to the earth's surface is about 18,000 m.p.h. and the time of orbit round the earth about 90 minutes; at 1,000 miles from the earth's surface the circular velocity is about 15,500 m.p.h. and time of orbit 2 hours; at 22,000 miles the time of orbit is one day (with interesting possibilities that have already been realized since the satellite can remain continually over one part of the earth[1]); while at 238,840 miles the circular velocity is only 2,300 m.p.h. and the time of orbit about 28 days, but there is no need to go to all the trouble of launching a satellite from this tremendous height—one has already been launched for us, the moon.

But how are we able to reach the moon and land on it?

[1] Examples are Early Bird and the original Telstar, which have been used as communication satellites to transmit television signals across the Atlantic.

Have a good look at Fig. 127; it will at least give you some idea. Remember too that, once a body is in orbit at a reasonable distance from the earth or the moon, there is no air resistance to slow it up, it continues automatically on its orbit (no power is needed), only a little thrust is required to change its path (and to halt the change), and it is weightless, as are

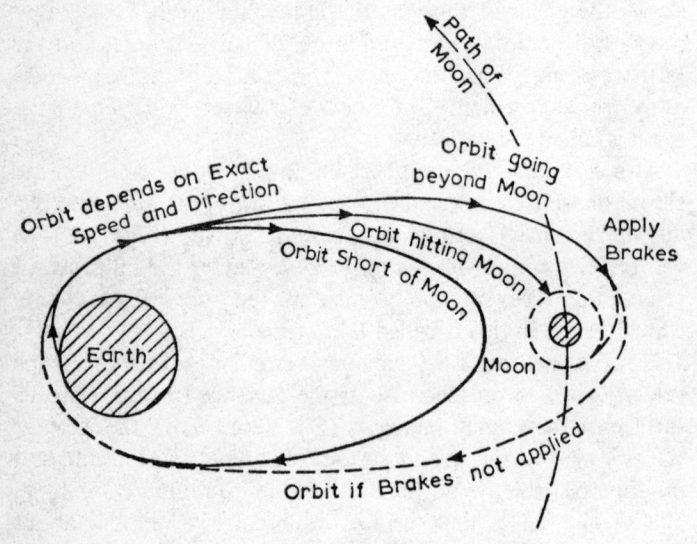

Fig. 127. **Sending satellites to the moon**
(Not to scale)

all things inside it (except during accelerations)—the diagram and a knowledge of these facts will help your understanding. So it is just a question of propelling a spaceship by a series of rockets to a suitable height, then launching it onto the right orbit, usually after some circular orbits of the earth, making any necessary adjustments to the course by firing auxiliary rockets (because great accuracy is needed), then applying the brakes (retrograde rockets) at the right time so as to get into

HAPPY LANDINGS

a circular orbit round the moon. (Incidentally, owing to the smaller mass of the moon the circular velocity will be less than a quarter of that round the earth—so fortunately will the escape velocity when we want to get away again!) While the spaceship circles the moon a part of it is detached to make a soft landing, again using rockets as brakes. After the occupants have seen enough of it, they take off (fairly easy with not much gravity), and join up with the main satellite—a feat practised on previous circuits round both the earth and the moon—then more rocket to escape from the attraction of the moon, and return towards earth in the usual way. Finally, and by no means the easiest problem, there is the re-entry of the earth's atmosphere, which has to be effected very accurately —both as regards angle and speed—to ensure that the craft does not bounce off it again, or burn up as it encounters the heat barrier. Once the drag and the increasingly dense atmosphere have reduced the speed to something reasonable, the spacecraft returns to earth, and splashdown, in a more or less normal parachute descent. That's all there is to it!—but we will probably have more to say about it in the next edition.

100. Happy Landings!

The author would indeed be happy if he could feel that the reader was as sorry as he is that this book is reaching its last pages. But this is asking too much. It is easy to write about flying, there is so much to write about, so many interesting things to discuss, so many knotty problems to be argued, so much that is obvious, yet wrong. It is a subject—one of the few, I believe—that it is easier to write about than to read about, easier to teach than to learn. But if the reader has been interested, that is the main thing; if he wants to find out more, if he wants to go and learn to fly, or, if he can already fly and wants to test the theory of flight in the air, then I shall be more than satisfied. If the reader feels that he has learnt little, then I shall be pleased, because my aim has been to make it all seem so simple that if the reader should learn anything he would do it unconsciously; there have been no mathematics, yet, at the same time, I have tried to cover most of the subject so far as it concerns the practical man, and I have tried too to be up to date and even to indicate development of the near future. I have tried—whether I have succeeded or not, only you can judge.

Next time you fly—and you must fly if you wish to understand the theory of flight—think of some of the things which you have read. You may not fly any the better for it, but you will be the more interested. And when you come down again —well, happy landings!

The Final Test

Before laying down the book, test yourself by answering the following questions. If you can do so, even if it is only to your own satisfaction, I feel that you have really learnt something of the subject, and you may either leave it at that or go on to read something a little more advanced. If you have difficulty in answering any of the questions you may turn back to the appropriate section, which is the same as the number of the question—but I hope that this may not be necessary.

1. Why is this book called *Flight Without Formulae*?
2. What is the correct name for:
    (*a*) An aeroplane that flies from the land only?
    (*b*) An aeroplane that flies from the water only?
    (*c*) An aeroplane that can fly from and alight on land or water?
    (*d*) Is a helicopter an aeroplane?

THE FINAL TEST

3. (*a*) Explain how an airship or balloon is kept in the air?
   (*b*) Why does an airship carry ballast?
4. Are airships vehicles of the past? If so, why?
5. How does (*a*) the temperature and (*b*) the density of the air change between sea-level and 50,000 ft? (*c*) Does the density change at the same rate as the pressure?
6. What is the difference between lift and drag?
7. (*a*) Distinguish between air speed and ground speed.
   (*b*) Must the wind be taken into account in handicapping an air race? If so, why?
8. How is it that a sailplane can gain height?
9. What is meant by scale effect?
10. What can be used as aids to "seeing the air?"
11. (*a*) Are we justified in watching how water flows in order to discover how air flows? If so, why?
    (*b*) What is a hydrofoil craft?
12. What do you understand by the terms (*a*) angle of attack, and (*b*) centre of pressure?
13. (*a*) What will make an aircraft tail-heavy?
    (*b*) What is meant by an unstable movement of the centre of pressure?
14. (*a*) Why are aeroplane wings usually cambered?
    (*b*) Are flat surfaces ever used as aeroplane wings?
15. What is downwash?
16. (*a*) What is meant by pressure plotting?
    (*b*) Show by a sketch how the pressure is distributed round a wing section at ordinary angles of flight.
17. What is a venturi tube? Give examples from ordinary life of the effects produced by a venturi tube.
18. How does the pressure distribution change as the angle of attack on a wing section is increased?
19. What is burbling?—and what causes it?
20. What is the correct meaning of "lift" as applied to aeroplanes?

21. What do you understand by the "speed squared" law?
22. What is (*a*) frontal area? (*b*) wetted surface?
23. How is drag affected by air density?
24. Explain the importance of a good lift/drag ratio.
25. What part of the drag has sometimes been given the flattering adjective *active*?
26. What is the cause of wing-tip vortices? What is induced drag?
27. What is parasite drag?
28. What part of the drag is reduced by streamlining? Is streamlining really effective? What is fineness ratio?
29. How is skin friction caused? Why is skin friction becoming of greater importance?
30. Explain what happens in the boundary layer.
31. (*a*) What are the effects of large camber on the upper surface of a wing?
    (*b*) Why has the tendency been towards a convex lower surface?
32. Is variable camber worth while?
33. (*a*) Describe the various types of flap. (*b*) What is the effect of slots?
34. What is aspect ratio, and what is its significance?
35. Why has the biplane proved the loser in its long struggle against the monoplane?
36. Sum up the problem of reducing the total drag of an aeroplane.
37. Why is straight and level flight an important consideration?
38. What are the four main forces acting on an aeroplane in flight?
39. On what principle is the thrust force provided in an aeroplane?
40. Describe jet propulsion. What is a ram-jet?
41. Can propeller propulsion and jet propulsion be combined?

THE FINAL TEST

42. What are (a) the advantages, (b) the disadvantages, of rocket propulsion?
43. How can the four main forces on an aeroplane be balanced in straight and level flight?
44. (a) Describe the function of a tail-plane. (b) What is a slab tail-plane?
45. What is meant by stability in an aeroplane?
46. Describe the various degrees of stability or instability.
47. What is meant by (a) yawing, (b) pitching, (c) rolling?
48. On what does longitudinal stability depend?
49. Explain how dihedral angle gives lateral stability.
50. What are the chief factors which influence directional stability?
51. Why are lateral and directional stability really inseparable?
52. On what does the effectiveness of a control surface depend?
53. How is longitudinal control made instinctive?
54. How is lateral control achieved?
55. Is the directional control of an aeroplane instinctive?
56. How may control surfaces be balanced?
57. What are control tabs? and for what purpose are they used?
58. (a) What is meant by aileron drag?
    (b) What are Frise ailerons?
    (c) What are differential ailerons?
    (d) What are spoilers?
59. What is meant by (a) flexibility, (b) elasticity, in an aeroplane structure? What is the purpose of mass balance?
60. What determines the speed range of an aeroplane?
61. (a) Why is economical flying important?
    (b) What is the difference between flying for maximum range and flying for maximum endurance?
    (c) How does jet propulsion influence the problem of economical flying?

62. Why is there a minimum speed below which flight is impossible?
63. What happens when an aeroplane is stalled?
64. What decides whether an aeroplane can land in a small space?
65. On what does the landing speed of an aeroplane depend?
66. What is meant by wing loading, and what is its significance?
67. (*a*) By what means can vertical take-off and landing be achieved?

    (*b*) What is the difference between a gyroplane and a helicopter?
68. (*a*) How is the gliding angle a measure of efficency?

    (*b*) Have you solved that problem about gliding to reach land?
69. Explain why the rate of climb decreases with height.
70. (*a*) On what does the correct angle of bank in a turn depend?

    (*b*) How can a pilot tell when the angle of bank is correct?

    (*c*) Does the function of the controls change in a steeply banked turn?

    (*d*) Discuss the possibility of a vertical bank.
71. Describe the forces acting on an aeroplane during a nose-dive.
72. (*a*) Why is a light aeroplane not very amenable to taxying?

    (*b*) Are the problems of taxying the same with a nose-wheel and a tail-wheel type of undercarriage?
73. Outline the process of taking off.
74. (*a*) What are the most common manoeuvres usually classed under the heading of aerobatics?

    (*b*) What is the state of affairs at the top of a good loop?

THE FINAL TEST

    (*c*) Explain how a spin starts.
    (*d*) What is the difference between a slow roll and a barrel roll?
    (*e*) Why is it difficult to do a steep side-slip on a modern aircraft?
    (*f*) What are the problems of inverted flight?
    (*g*) What are the causes of bumpy weather?

75. (*a*) Explain the difference between the pitch and the advance per revolution of a propeller.
    (*b*) Why does the blade angle of a propeller decrease from boss to tip?
    (*c*) What are the advantages of variable pitch in a propeller?
    What is a constant speed propeller? What is feathering? Describe two ways in which a propeller can be used as a brake.
    (*d*) What is the slipstream, and what are its effects?

76. What are the advantages of multi-engined aeroplanes?

77. (*a*) Explain how trimming tabs may be used to correct flying faults.
    (*b*) What are the possible causes of an aeroplane flying (i) left wing low, (ii) nose-heavy, (iii) with a tendency to turn to the right?

78. What are the limitations of flying *by the seat of one's pants*?

79. What is the difference between true and indicated speed?

80. What does an altimeter measure?

81. What is meant by (*a*) deviation, (*b*) variation, in a compass reading?

82. What is the main purpose of the following instruments: (*a*) artificial horizon, (*b*) directional gyro, (*c*) turn and and side-slip indicator?

83. What has the speed of sound got to do with the problem of high-speed flight?

84. At what speed does sound travel in air? Is this a constant value?
85. What is meant by a Mach number?
86. Why do we make a special study of flight at transonic speeds?
87. What is a shock wave?
88. What is a shock stall? What are the similarities and differences between a shock stall and an ordinary stall?
89. How and why do shock waves create extra drag?
90. What is the advantage of sweepback at transonic speeds?
91. What is the purpose of vortex generators?
92. (*a*) What shapes of wing and body are most suitable for transonic speeds?
    (*b*) What is the area rule?
93. How does the pattern of shock waves change as we pass through the sound barrier?
94. What are the differences between air flowing at supersonic and subsonic speeds?
95. What are good wing shapes for supersonic speeds?
96. What is the cause of sonic bangs?
97. Are there any limits to flight at supersonic speeds? What is meant by "variable geometry"?
98. What are hypersonic speeds?
99. What decides whether a missile will (*a*) come back to earth, (*b*) orbit the earth, (*c*) escape into space? (*d*) Describe the problems that were faced by man in landing on the moon.
100. What have you learnt from this book? Do you want to learn more? If so, then I suggest that when you fly, whether as pilot or passenger, you keep your eyes open and notice how the things that have been described in this book *actually happen*—maybe you will be surprised that they do. If you want to learn still more, then you must turn to other books; there are plenty of them, too

THE FINAL TEST

many in fact, and you must choose with discretion and on the advice of those who know. Many of the best of these books require some knowledge of mathematics and mechanics, but you must remember that there is a limit to what you can expect to learn "without formulae."

Oh, by the way, do you remember the official definition of an aeroplane given in one of the first paragraphs of the book? To save you looking back, here it is again. It is still a bit of a mouthful, but I hope that it may now convey to you a great deal more than it did when you first read it.

An aeroplane is a heavier-than-air flying machine, supported by aerofoils, designed to obtain, when driven through the air at an angle inclined to the direction of motion, a reaction from the air approximately at right angles to their surfaces.

Yes—that's what an aeroplane is—but it is much more than that, isn't it?

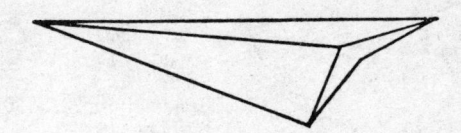

Fig. 128. **A paper dart—the shape of the future?**

INDEX

(*References are to page numbers*)

ABSOLUTE ceiling, 164
Aerobatics, 186–96
Aerofoil, 34–40
Aeroplane—
  (definition), 1–4
  parts of, 5
Aileron drag, 122
Ailerons, 114
  differential, 124
  Frise, 125
Air brakes, 142, 157
Aircraft (definition), 2, 4
Air density, effects of, 53, 54
Air flow, 35–7
Airscrew, 197
Airships, 4, 7–12
Air speed, 19–22
  indicated, 217
  indicator, 215–18
  true, 217
Altimeter, 218–20
Angle—
  blade, 199
  dihedral, 109
  of attack, 31, 46, 47
  of incidence, 31
  pitch, 199
Apogee, 258, 261
Archimedes' Principle, 6
Area rule, 243
Artificial horizon, 224
Aspect ratio, 77–86
Atmosphere, 12–17
  International Standard, 15, 16, 219
Attack, angle of, 31, 46, 47
Autogiro, 146
Auto-rotation, 189

Axis—
  lateral, 104–6
  longitudinal, 104–6
  normal, 104–6

BALANCE, 87, 94–8
  mass, 129, 130
  tabs, 121
Balanced controls, 116–19
Balloons, 4, 7–12
Barrier—
  heat, 253
  sonic, 228 *et seq.*
Bernoulli's Theorem, 43
Biplane interference, 81
Biplanes, 80–3
Blade angle, 199
Boundary layer, 67–72
Brakes, air, 142, 157
Braking (propeller), 202
Burbling, 46–9

CAMBER, variable, 73, 74
Ceiling—
  absolute, 164
  service, 164
Centre of pressure, 31, 39, 45
Centrifugal force, 171
Chord line, 46, 47
Circular velocity, 261
Circulation, 70
Climbing, 163–70
Compass, 220–2
Compressibility, 30, 226 *et seq.*
Control, 112–31
  at high speeds, 127–31
  at low speeds, 122–7
  directional, 115

272

INDEX

Control (*contd.*)—
  lateral, 114
  longitudinal, 114
  tabs, 119–22
Controls—
  balanced, 116–19
  irreversible, 130

DELTA wing, 242
Density, 13 *et seq.*
Deviation, 222
Differential ailerons, 124
Dihedral angle, 109
Directional—
  control, 115
  gyro, 224
  stability, 110–12
Distribution, pressure, 37–40
Drag, 17–19, 49, 50, 84, 85
  active, 57
  aileron, 122
  analysis of, 55–7
  cooling, 67
  form, 56, 62–5
  induced, 56, 57–60, 78
  interference, 65
  parasite, 56, 60–2
  shock, 238
  wave, 235 *et seq.*
  wing, 56, 57
Dynamic pressure, 43 *et seq.*

ECONOMICAL flying, 134–7
  with jets, 136
Elevators, 99, 114
Endurance, flying, 135
Equilibrium, 32
Escape velocity, 261

FEATHERING (propeller), 202
Fineness ratio, 65
Flaps, 74–7, 143, 144
Flutter, 129
Flying faults, 206–13
Form drag, 56, 62–5
Frise ailerons, 125

Frontal area, 51–3

GLIDING, 150–63
Ground—
  effect, 140
  speed, 19–22

HEAT barrier, 253
Helicopter, 34, 148
High-speed flight, 226 *et seq.*
Hovercraft, 3, 140
Hydrofoil craft, 31
Hypersonic speeds, 255

INCIDENCE, angle of, 31
Indicated air speed, 217
Indicator, air-speed, 215 *et seq.*
  turn and side-slip, 223
Induced drag, 56, 57–60, 78
Instability, spiral, 112
Instruments, 213 *et seq.*
  Air-speed indicator, 215–18
  Altimeter, 218–20
  Artificial horizon, 224
  Compass, 220–2
  Directional gyro, 224
  "George," 225
  Mach meter, 225
  Turn and side-slip indicator, 225
International Standard Atmosphere, 15, 16, 219
Irreversible controls, 130

JET propulsion, 89, 90

LAMINAR flow, 73
Landing, 139–43
  speed, 139–45
Lateral—
  axis, 104 *et seq.*
  control, 114
  stability, 108–10
Layer, boundary, 67–72
Lift, 17–19, 49, 50, 84, 85
Lift/drag ratio, 54, 55, 153

Lighter-than-air, 3–12
Loading, wing, 145, 146
Longitudinal—
  axis, 104 *et seq.*
  control, 114
  stability, 106–8
Looping, 187

MACH number, 229 *et seq.*
Manœuvres—
  Aerobatics, 186–96
  Climbing, 163–70
  Gliding, 150–63
  Landing, 139–50
  Looping, 187
  Nose-diving, 180–3
  Rolling, 191
  Side-slipping, 192
  Spinning, 189
  Stalling, 137–9
  Straight and level flight, 85, 86
  Taking-off, 184–6
  Taxying, 183, 184
  Turning, 171–80
  Upside-down-flight, 193
Man-powered aircraft, 2, 4
Mass balance, 129, 130
Missiles, 256 *et seq.*
Moon landing, 261–3
Multi-engine aeroplanes, 205, 206

NORMAL axis, 104–6
Nose-diving, 180–3
Nose-heavy, 94
Number—
  Mach, 229 *et seq.*
  Reynolds, 26

PARASITE drag, 56, 60–2
Perigee, 258, 261
Pitch (of propeller), 197
  angle, 199
  reversible, 202
  variable, 201
Pitching, 104, 106
Pitot-static head, 216

Plane, tail, 98–100
Pressure—
  atmospheric, 14
  centre of, 31, 39, 45
  distribution, 37–40
  dynamic, 43 *et seq.*
  plotting, 38
  static, 43 *et seq.*
Propeller, 196–204
  braking, 202
  feathering, 202
  propulsion, 90–2
Propulsion—
  jet, 89, 90
  propeller, 90–2
  rocket, 92–4
Pusher propeller, 198

RAM-JET, 89, 90
Range and endurance, 134–7
Reversible pitch, 202
Reynolds number, 26
Rocket propulsion, 92–4
Rolling, 104–6, 191
Rotorcraft, 3, 4, 147 *et seq.*
Rudder, 115

SATELLITES, 256 *et seq.*
Scale effect, 26
Service ceiling, 164
Sesquiplane, 83
Shape of wing section, 34, 72, 249
Shock drag, 238
Shock stall, 232 *et seq.*
Shock waves, 232 *et seq.*
Side-slipping, 192, 193
Size, effects of, 51
Skin friction, 52, 56, 65–7
Slab tail-plane, 100
Slats, 76
Slipstream, 203
Slots, 74–7, 126
Smoke tunnels, 28, 29
Sonic bangs, 251, 252
Sonic barrier, 228 *et seq.*
Sound, speed of, 30, 226 *et seq.*

INDEX

Space, 256 et seq.
Speed—
　air, 19–22
　effects of, 50
　ground, 19–22
　hypersonic, 255
　landing, 139–45
　of sound, 30, 226 et seq.
　sonic, 30, 226 et seq.
　stalling, 138
　subsonic, 230
　supersonic, 230, 243 et seq.
　transonic, 230 et seq.
Speed-range, 131 et seq.
Spinning, 189
Spiral instability, 112
Spoilers, 126
Spring tab, 121
Stability, 32, 33, 100 et seq.
　automatic, 104
　degrees of, 103, 104
　directional, 110–12
　inherent, 104
　lateral, 108–10
　longitudinal, 106–8
Stabilizer, 98
Stagger, 81–3
Stall, shock, 232 et seq.
Stalling, 46–9, 137–9
　speed, 138
Static pressure, 43 et seq.
S.T.O.L., 146–9
Straight and level flight, 85, 86
Stratosphere, 15
Streamlining, 62 et seq.
Supersonic—
　flow, 247 et seq.
　shapes, 248 et seq.
　speed, 230, 243 et seq.
Surface area, 52
Sweepback, 238 et seq.

TAB—
　balance, 121
　spring, 121
　trimming, 121

Tabs, control, 119–22
Tail-heavy, 94
Tail plane, 98–100
Taking off, 184–6
Taxying, 183–4
Temperature, atmospheric, 14
Terminal velocity, 181
Thrust, 87, 88, 89
　of propeller, 198
Torque, of propeller, 198
Tractor (propeller), 198
Transonic speeds, 230, 231 et seq.
Tricycle undercarriage, 140
Trimming tabs, 121
Tropopause, 15
Troposphere, 15
True air speed, 217
Tube, venturi, 40–5
Tunnels—
　smoke, 28, 29
　wind, 23–8
Turbo-jet, 90
Turbo-prop, 91
Turn and sideslip indicator, 225
Turning, 171–80

UNDERCARRIAGE, tricycle, 140
Upside-down-flight, 193
VARIABLE—
　camber, 73, 74
　geometry, 254
　pitch, 201
Variation, 222
Velocity—
　circular, 261
　escape, 261
　terminal, 181
Venturi tube, 40–5
Viscosity, 68
Vortex generators, 240, 241
Vortices, wing-tip, 57–60
V.T.O.L., 146–50

WAVES, shock, 232 et seq.
Weight, 87

Wind—
   gradient, 159
   tunnels, 23-8
Wing—
   loading, 145, 146

Wing (*contd.*)—
   section, 34-40, 72, 73
   -tip vortices, 57-60

YAWING, 104-6

LIST OF PLATES

Every picture tells a story. The plates that follow have been carefully chosen to tell the story of Flight Without Formulae. To those who have read the book they will tell the story again in a new form and thus provide an excellent means of revision. It is hoped, of course, that the reader will have referred to them while studying the text, but even if he has he should look through them again, and read the notes underneath each picture, because the book is not complete without them, and they have purposely been put at the end of the book to form, as it were, a closing chapter. The plates have been arranged in two groups; Group 1 (Plates 1-32) illustrates the history of heavier-than-air flight from 1903 to the present day, and Group 2 (Plates 33-64) shows various types of flying machine, both old and new, with even a glimpse of the future, and the devices used to propel, control and manoeuvre them.

Group 1

1. The early days. 1903. The Wright biplane
2. The early days. The Bleriot monoplane
3. First World War. The B.E.2C
4. First World War. The S.E.5A
5. First World War. The Handley Page 0/400
6. First World War. The Bristol triplane

7. Between the wars. The Bristol Bulldog
8. Between the wars. The Gloster Gladiator
9. Between the wars. The Westland Lysander
10. Between the wars. The De Havilland Albatross
11. Second World War. The Vickers Wellington
12. Second World War. The De Havilland Tiger Moth
13. Second World War. The Avro Lancaster
14. Second World War. The Hawker Hurricane
15. Second World War. The De Havilland Mosquito
16. Second World War. The Gloster Meteor
17. After the wars. The De Havilland 108
18. After the wars. The Blackburn Beverley
19. After the wars. The Bristol Britannia
20. After the wars. The Avro Vulcan
21. After the wars. The Gloster Javelin
22. After the wars. The English Electric Lightning
23. After the wars. The Fairey Delta 2
24. Flight today. The De Havilland Comet 4
25. Flight today. The Vickers VC.10
26. Flight today. The Boeing 707
27. Flight today. The BAC One-Eleven
28. Flight today. General Dynamics F111
29. Flight today. The Boeing 727
30. Flight today. The Britten-Norman Islander
31. Flight today. The Anglo-French Jaguar
32. Flight today. The Anglo-French Concorde

Group 2

33. Lighter than air. A captive or kite balloon
34. Lighter than air. Free balloons
35. Lighter than air. An airship. The Graf Zeppelin
36. A float plane. The Fairey Swordfish
37. A flying boat. The Saunders-Roe S.R.A.1

38. An Autogiro. The Cierva C.24
39. A helicopter. The Westland W.G.13
40. Helicopter-cum-autogyro. The Fairey Rotodyne
41. New type rotorcraft. The Westland WE-02
42. S.T.O.L., old style. Scottish Aviation Twin Pioneer
43. V.T.O.L., old style. The Short S.C.1
44. V./S.T.O.L., new style. The Hawker Siddeley Harrier
45. V.T.O. The Saturn V moon rocket
46. Contrasts. A sailplane. The Slingsby Skylark 4
47. Contrasts. A rocket. The Black Knight
48. Contrasts. A hovercraft. The B.H.C. SRN.6
49. Contrasts. A hydrofoil craft. Gibbs & Cox Sea Legs
50. Trainer, old style. The Avro 504 K
51. Trainer, new style. The Hunting Jet Provost
52. Pusher, old style. The Maurice Farman Shorthorn
53. Pusher, new style. The BAC One-Eleven
54. Parasite drag. The Vickers Virginia
55. Streamlining. The English Electric Canberra
56. Thrust for flight. 1914. Propeller and rotary engine
57. Thrust for flight. Turbo-props
58. Thrust for flight. Turbo-jets and rocket boosters
59. Thrust for flight. One man-power
60. Putting the brakes on. The English Electric Lightning
61. Putting the Brakes on. The Handley Page Victor
62. What might have been. The TSR2
63. Another "might have been"? The Boeing supersonic transport
64. Anglo-French supersonic transport. The Concorde

Group 1: Plates 1–32

PLATE 1. THE EARLY DAYS. 1903. THE WRIGHT BIPLANE
(*By courtesy of "Flight"*)

A contraption of struts and wires, flying tail first, landing on skids. The two propellers were driven by chains from a single engine. Unstable and very nearly unmanageable—but it flew!

PLATE 2. THE EARLY DAYS. THE BLERIOT MONOPLANE
(*By courtesy of "Flight"*)

Made first flight across the English Channel on the 25th July, 1909, at an average speed of 36 m.p.h. Although this is a monoplane, it differs from modern types in that it is braced top and bottom by landing and flying wires. It took twenty years to get rid of the external bracing of monoplanes.

PLATE 3. FIRST WORLD WAR. THE B.E.2C

(*By courtesy of "Flight"*)

Product of the Royal Aircraft Factory at Farnborough, but designed by the famous De Havilland family. The B.E.2C owed its fame, or notoriety, to its war-time history and to its rather peculiar flying habits. It had a way of its own—a certain inherent stability—which pilots of those days (the author among them) did not like. It had an air-cooled vee-type in-line engine—an unusual combination.

PLATE 4. FIRST WORLD WAR. THE S.E.5A
(By courtesy of "Flight")

Famous fighting scout of conventional design, except that it had a much higher degree of stability than was then considered advisable for a fighter. Another product of the Royal Aircraft Factory, it bore some resemblance to the B.E2C and also to later products of the Gloster Company, on which the same designer was at work. Fitted with a water-cooled engine.

PLATE 5. FIRST WORLD WAR. THE HANDLEY PAGE 0/400
(By courtesy of Handley Page Ltd.)

This forerunner of all large bombers was extensively used during the First World War. The structure was typical of large machines of the day, being fitted with a considerable extension of the top planes over the bottom.

PLATE 6. FIRST WORLD WAR. THE BRISTOL TRIPLANE
(By courtesy of the Bristol Aeroplane Co. Ltd.)

Thirty of these were ready to bomb Berlin when the armistice was signed in 1918. They represent the familiar wire-braced frame applied to a triplane, and some people thought that the large aircraft of the future would be on these lines, but how wrong they were! Notice that even the tail unit has become a biplane on both these large bombers.

PLATE 7. BETWEEN THE WARS. THE BRISTOL BULLDOG

(By courtesy of the Bristol Aeroplane Co. Ltd.)

One of the most famous representatives of the typical biplane structure, but notice that, even on a fairly small aircraft the top plane has grown in proportion to the lower; we were already trying to get the advantages of a monoplane. This machine may be considered as typical of those that came between the two wars. Notice the spinner on the propeller and the engine cowling—both new features.

PLATE 8. BETWEEN THE WARS. THE GLOSTER GLADIATOR
(*By courtesy of "Flight"*)

The last, and perhaps the best, of all great biplane fighting aircraft. Notice the spinner again, the engine cowling encircling the whole engine, the enclosed cockpit, the tail-wheel instead of skid, and the simplified undercarriage. Things were changing!

PLATE 9. BETWEEN THE WARS. THE WESTLAND LYSANDER
(*By courtesy of "Flight"*)

Things *have* changed. A monoplane again, with slots and flaps for the first time. The undercarriage is still not retractable, but the struts are well faired, and both main and tail wheels are enclosed in "spats." This was a freak aeroplane designed to fly slowly for observation purposes, but it is typical of experiments and progress between the two wars.

PLATE 10. BETWEEN THE WARS. THE DE HAVILLAND ALBATROSS
(By courtesy of "Flight")

Contrast this with the ideas of a large aeroplane conveyed by Plates 5 and 6. Here we have a roomy fuselage, excellent streamlining, a monoplane structure without any external wires or struts, a retractable undercarriage, engines in the wings. These improvements represented the progress of some twenty years.

PLATE 11. SECOND WORLD WAR. THE VICKERS WELLINGTON

(By courtesy of "Flight")

This famous twin-engined bomber differed in construction from the Battles and Blenheims that were much used in the early days of the Second World War in that it had a lattice work "geodetic" framework with fabric covering. This light construction, compared with the standard cantilever monoplane with metal covering that was so characteristic of both fighters and bombers at the time made it possible to have a large wing span and high aspect ratio thus reducing the induced drag. The same type of construction was used on the single-engined Wellesleys which gained the world's long-distance record in a formation flight to Australia before the Second World War.

PLATE 12. SECOND WORLD WAR. THE DE HAVILLAND
TIGER MOTH

(*By courtesy of "Flight"*)

One of the standard elementary trainers of the Second World War, but always a little suspect in being a biplane. The Tiger Moth had its rivals—there was the little Magister and, overseas, the Cornell, and for advanced training the Master and the Harvard; for twin-engine training the Oxford and the Cessna. All of these were monoplanes. But the Tiger Moth, perhaps the last real biplane, never lost the affection of those who believed in it.

PLATE 13. SECOND WORLD WAR. THE AVRO LANCASTER

(*By courtesy of "The Aeroplane"*)

One of the outstanding four-engined bombers of the Second World War. This, and its brother heavy bombers, the Stirling and Halifax and the corresponding American types, illustrate the difficulty of reconciling operational needs with clean lines and streamlining—compare, for instance, with the civil Albatross of Plate 10. In spite of their looks, however, they achieved a remarkable performance in the weight that they could carry and the distance over which they carried it.

PLATE 14. SECOND WORLD WAR. THE HAWKER HURRICANE
(*By courtesy of "Flight"*)

The Hurricane and its friendly rival the Spitfire were the outstanding British fighter aircraft of the Second World War. Designed before the war, they survived to the end, and the Spitfire was still on active service in Malaya in 1950. Note the tail-down attitude that is a characteristic of inverted flight.

PLATE 15. SECOND WORLD WAR. THE DE HAVILLAND MOSQUITO

(By courtesy of "The Aeroplane")

The most remarkable aircraft of the Second World War, and at one time the fastest military aircraft in the world. Remarkable too in its versatility—fighter, bomber and reconnaissance type all merged into one. Look carefully at its lines and you will see one of the main reasons for its success. But what you will not see is that, unlike other aircraft of its time (except the Oxford trainer), it was of wooden construction; this was largely responsible for its lightness and remarkable performance.

PLATE 16. SECOND WORLD WAR. THE GLOSTER METEOR
(*By courtesy of the Gloster Aircraft Co. Ltd.*)

The only British jet-driven aircraft to become operational during the war. After the war the Meteor soon came into its own. The absence of propellers enabled a low undercarriage to be used, thus saving weight. The nose-wheel undercarriage enabled the aircraft to stand in the flying attitude on the ground, and so the hot jets did not strike the runway. Note, too, the high position of the tailplane which had to be kept clear of the jets.

PLATE 17. AFTER THE WARS. THE DE HAVILLAND 108
(*By courtesy of the De Havilland Aircraft Co. Ltd.*)

After the wars there was a period of experiment with all types of aircraft, large and small, fast and slow, civil and military.

The Bristol Brabazon had eight engines and a minimum weight of 130 tons; here, in contrast, is the D.H. 108, which, flying soon after the Second World War, represented a landmark in this history in pictures of the progress of flight. Contrast it with the Wright biplane of 1903, and you will see what I mean. Biplane has become monoplane; all external bracing and encumbrances have disappeared, so have the tails; wings have been swept back and the body streamlined; there are flaps and air brakes and even slots—though we can't see them in the picture; and finally, propellers, chains and reciprocating engine have been replaced by a jet. These are the things that we have learnt about flight in the intervening period, and the gradual change can be traced through the types that we have illustrated. The speed of the Wright biplane was about 35 m.p.h. In April, 1948, a D.H. 108 flew at 605 m.p.h.

PLATE 18. AFTER THE WARS. THE BLACKBURN BEVERLEY

(*By courtesy of Blackburn and General Aircraft, Ltd.*)

The freighter too made progress, though in a different direction from that of the De Havilland 108. First there was the Bristol Freighter, which could carry awkward shaped loads at 220 m.p.h. Then the Beverley which could hardly be called beautiful and it may have deserved some of its nicknames, but it had space for 7,500 cu ft of freight and could carry 22 tons of it, or 160 troops, or various combinations of both—and all at speeds up to 240 m.p.h. Its cargoes included helicopters (two at a time), bulldozers, mechanical shovels, electric generators, mobile factories, and single mites of equipment up to 17 tons in weight and 50 ft in length. Contrast this with the Wright machine, which like all the early types, had great difficulty in lifting itself.

PLATE 19. AFTER THE WARS. THE BRISTOL BRITANNIA

(*By courtesy of the Bristol Aeroplane Co. Ltd.*)

And here is progress—in yet another direction. The Britannia shows that progress and beauty can go hand in hand, when all we want to do is to carry a reasonable number of passengers (say 100), at reasonably high speeds (say 400 m.p.h.), over reasonable distances (say 4,000 miles). The Britannia came from the same stable as the Brabazon, but in the meantime the lesson of sweet reasonableness had been learnt; the Brabazon proved to be too big and too heavy for commercial operation; the Britannia proved ideal. Propulsion was by four turbo-props.

PLATE 20. AFTER THE WARS. THE AVRO VULCAN

(*By courtesy of A. V. Roe & Co. Ltd.*)

Here is the world's first four-jet delta wing bomber, one of the famous trio of V-bombers (which included the Vickers Valiant and the Handley Page Victor) supplied to the Royal Air Force soon after the end of the war. The specification demanded more than 100 per cent advance in performance—speed, altitude, range and load-carrying capacity—and the Vulcan more than fulfilled the requirements. Many long-distance flights were accomplished at average speeds around 625 m.p.h. Notice the clean shape of the delta wing completely concealing all but the intake and jet pipes of the four engines, each of which gave a thrust of more than 10,000 lb in the early types; this was increased by later developments up to as much as 17,000 lb each. The beautiful proportions disguise the real size of the aircraft, which in this version has a span and length of just under 100 ft.

The three V-bombers were designed to the same specification, and this makes it all the more interesting to note their similarities and differences; unfortunately the space at our disposal does not allow us to illustrate all three. Both the Victor and the Valiant had conventional tail units instead of the delta wing; both had wings of crescent shape with more sweepback inboard than outboard, thus ensuring a high critical Mach number combined with good characteristics at low speed.

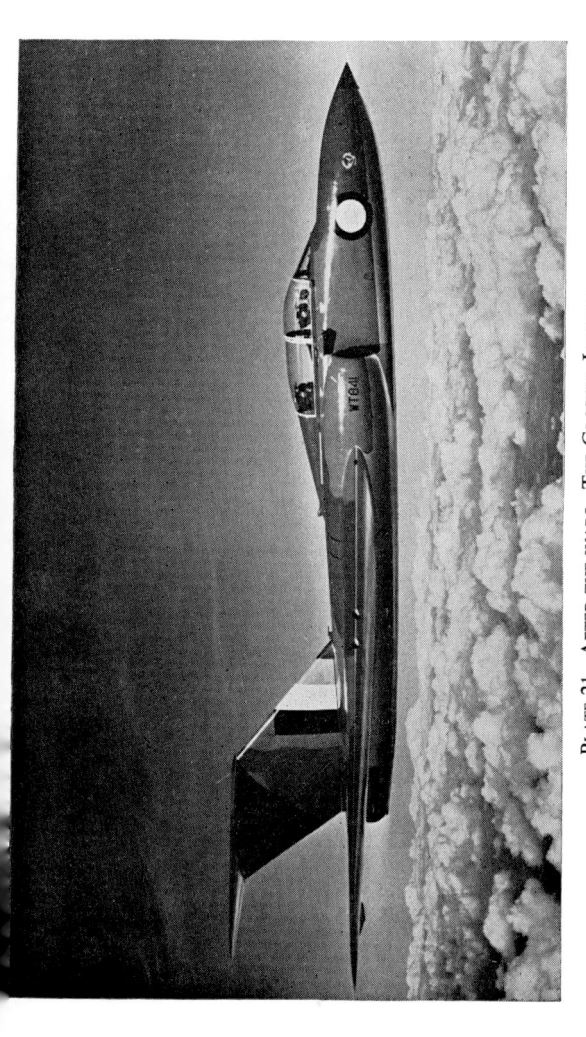

PLATE 21. AFTER THE WARS. THE GLOSTER JAVELIN

(By courtesy of the Gloster Aircraft Co. Ltd.)

In this and the next plate we see the corresponding development in fighter aircraft. The prototype of the Javelin first flew in 1951, and there were at least seven production versions. It is perhaps the last fighter from a famous fighting stable, and this picture prompts us to look back at the Meteor (Plate 16), and the Gladiator (Plate 8), and perhaps even at the S.E.5A (Plate 4), to see whether we can trace any family likeness running through from start to finish. In the Meteor we noticed the high position of the tail-plane, but that was nothing compared with the Javelin, where it has gone to the limit at the top of an exceptionally large fin and rudder. The later versions had an all-moving, or slab, tail-plane with elevators virtually acting as tabs. Since the Javelin was very much a product of the transonic era, various devices were tried to reduce shock and flow separation; to this end most of the production versions had vortex generators on the top surface of the wings and thickened trailing edges, which had a similar effect, at the rear of the wings.

PLATE 22. AFTER THE WARS. THE ENGLISH ELECTRIC LIGHTNING

(*By courtesy of the English Electric Co. Ltd.*)

This was to be the last manned fighter of the Royal Air Force, and as such it would have been a worthy climax to a wonderful story not only of fighting in the air but of progress in aircraft design. Here we have sweepback in the extreme, both in wings and tail, 60° of it, and this and the clean lines and more than 11,000 lb of thrust from each of the two engines (one above the other) give the clue to a speed of over 1,200 m.p.h. There is also provision for fitting a rocket engine in place of the fuel tanks below the fuselage. But not the least interesting feature of this picture is that the Lightning is carrying two guided missiles (De Havilland Firestreaks), early examples perhaps of the aircraft of the future, but with their long noses and small span wings and tails not so very unlike the latest developments in present-day manned aircraft such as the Lightning itself. What we can't see in this picture is how an aircraft capable of such high speeds can land and pull up in less than 900 yards; the secret of this is revealed in Plates 60 and 61.

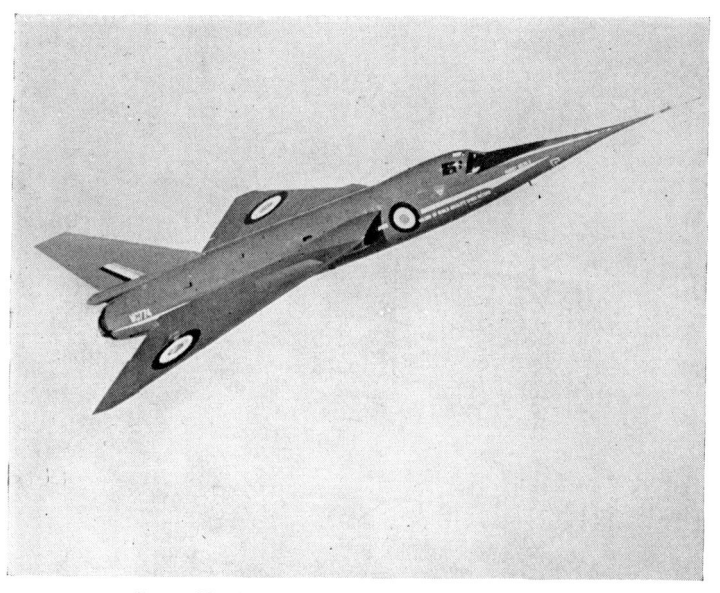

PLATE 23. AFTER THE WARS. THE FAIREY DELTA 2
(*By courtesy of the Fairey Aviation Co. Ltd.*)

The Fairey Delta 2, primarily a research aircraft, first flew as long ago as 1954, but it has been selected as a fitting landmark in this story in pictures of the progress in aeroplane design during over 60 years of flight, and in particular during the experimental period after the wars. The reason for the choice will be clear if we look back to the Wright biplane in Plate 1. Here are two aircraft which appear to be flying in much the same attitude in much the same direction, yet could there be a greater contrast?—whether in shape, in construction, in means of propulsion, or in the position of the pilot's legs? In describing the De Havilland 108 (Plate 17) as a landmark in progress immediately after the Second World War, its speed of 605 m.p.h., achieved in 1948, was contrasted with the 35 m.p.h. of the Wright biplane. In March, 1956, only eight years later, the F.D.2 flew at 1,132 m.p.h., setting up a world's speed record which stood for 18 months, when the U.S.A. put it up to 1,404 m.p.h.

PLATE 24. FLIGHT TODAY. THE DE HAVILLAND COMET 4

(*By courtesy of the De Havilland Aircraft Co. Ltd.*)

The De Havilland Comet 1 was the first real commercial jet air liner, and such was its success from the operational point of view that the firm was undaunted by the severe setback caused by more than one structural disaster. After the cause of the failures had been investigated they set to work not only to prevent any similar failure in future designs but to improve on performance in all directions—and here is the result, the Comet 4. Notice the clean design, the long slim body, the long nose, the low position of the wing (so that the jets clear the tail); notice, too, how the jet engines have been placed close together and close to the body, leaving a large portion of the wing of true aerofoil shape, and reducing the turning effect when one engine is out of action. This is something that cannot be done when propellers are used. But it is interesting to note that in the closest rival of the Comet 4, the American Boeing 707 (Plate 26), the jet engines are very widely spaced, though the wings are kept clean by placing the jets below and well clear of the wings.

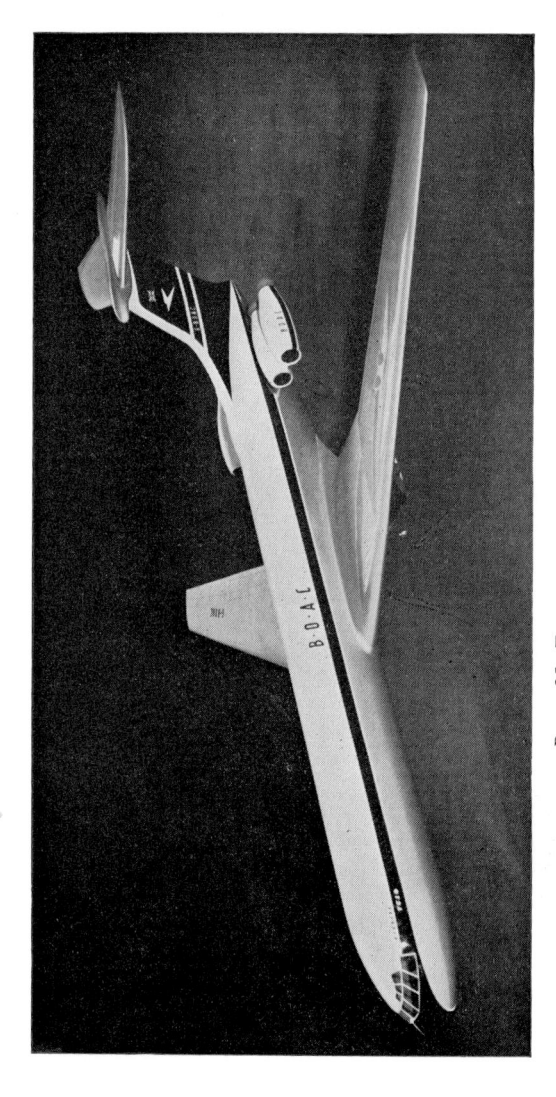

PLATE 25. FLIGHT TODAY. THE VICKERS VC.10

(By courtesy of Vickers-Armstrongs (Aircraft) Ltd.)

When one looks at the general lines of the Britannia and the Comet 4, it is not easy to see what future progress there can be in the shape of commercial air liners—so long, at least, as we are content with high subsonic speeds in the region of 500 to 600 m.p.h. But here in the VC.10 is one possible line of progress—putting the engines at the tail, a real pusher again. Notice the beautifully clean wing, placed even farther back with the weight of the engines so far to the rear; notice the even higher tail, and imagine the increased comfort to the passengers from the lack of engine noise and vibration. As so often happens, this idea of putting the jet nacelles at the rear of the fuselage was developed simultaneously in several different countries. At maximum take-off weight the wing loading of the VC.10 exceeds 100 lb/sq ft.

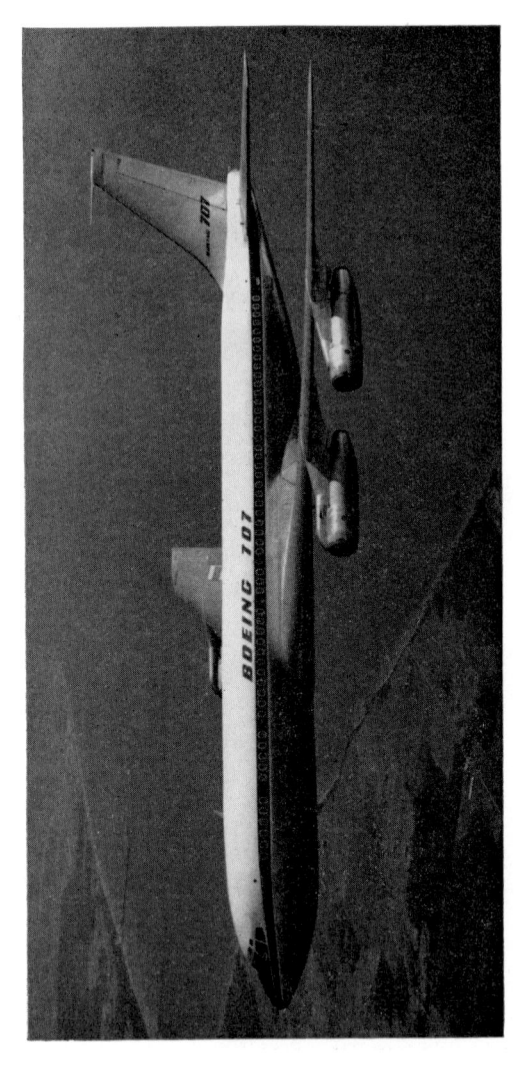

PLATE 26. FLIGHT TODAY. THE BOEING 707

(*By courtesy of the Boeing Company*)

The Boeing 707 is described as an intercontinental multi-purpose jet. Intercontinental it is indeed with a range of more than 3,500 miles, and multi-purpose in that it was specifically designed for carrying either passengers or cargo, or a combination of the two; nearly 200 passengers or a cargo payload of 45 tons, at speeds of over 600 m.p.h. In view of its performance it is hardly surprising that, with the exception of the D.C.3, there are more of these aircraft flying today than any other long-distance airliner.

Note that the four turbofan engines, each of 18,000 lb thrust, are well below the wings so that the jet streams clear the tail, which can thus be placed on the centre-line of the fuselage instead of in the rather awkward position at the top of the fin as in the majority of jet aircraft.

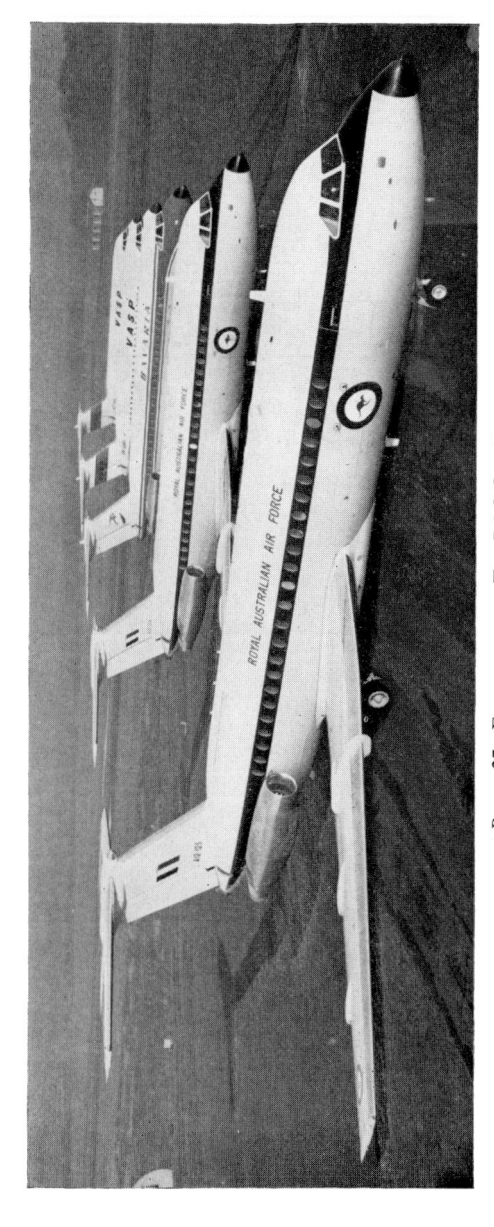

PLATE 27. FLIGHT TODAY. THE BAC ONE-ELEVEN

(*By courtesy of the British Aircraft Corporation*)

This picture of a flight line-up of five BAC 1-11s gives some idea of their world-wide popularity and success. These aircraft have proved ideal for internal airlines and are now operating with 20 such lines in 40 countries, and serving more than 130 cities. Notice the fences on the wing, the high tail-plane and the engines at the rear. A new version, the Super 1-11, is longer and with increased engine thrust, can carry more passengers and cargo with even better performance.

PLATE 28. FLIGHT TODAY. THE GENERAL DYNAMICS F111

(*By courtesy of General Dynamics, Fort Worth, U.S.A.*)

This, the first successful operational swing-wing aircraft of American design, but based on the original idea of Mr. Barnes Wallis, has had its teething troubles (which is hardly surprising in such a novel design) and has been a subject of controversy both in this country and in the United States. But as a military aircraft its performance is remarkable; it is a two-man, all-weather, multi-purpose, supersonic fighter-bomber capable of carrying both conventional and nuclear weapons.

The upper picture shows the wings at the 16° sweepback position for take-off, and the lower one at the full 72·5° position for supersonic speeds up to 60,000 ft. Intermediate positions can be used for cruising.

After the cancellation of the TSR2 (Plate 62), 50 F111s were ordered for the Royal Air Force, and their cancellation means that the British Services will be left without a long-range strike/reconnaissance aircraft when the V-bombers and the Canberra become obsolete.

But the British may yet come into their own with a smaller and cheaper swing-wing type?

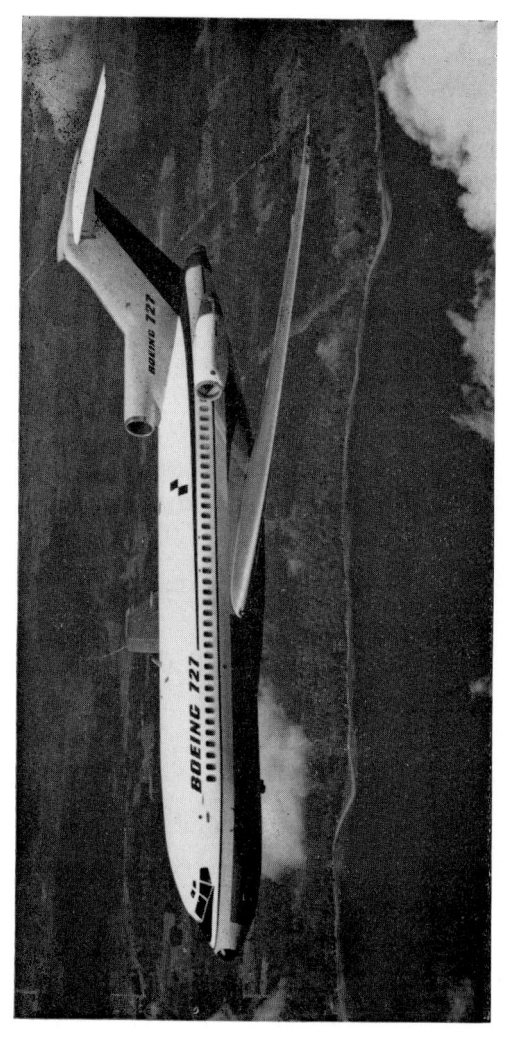

PLATE 29. FLIGHT TODAY. THE BOEING 727

(*By courtesy of the Boeing Company*)

Compare this with the 707 in Plate 26. Here we have three engines, each of 14,000 lb thrust, instead of four, and they are at the rear instead of under the wings, and so there is the high tail-plane. But, as in the whole series of Boeing airliners, there are remarkable resemblances and many interchangeable parts, the differences being simply due to the different roles for which the aircraft are designed. The purpose of the 727 is to bring jet speed, comfort and reliability to short-to-medium ranges (150 to 1,700 miles), but the speed of about 600 m.p.h. is similar to that of the 707, and the fuselage, though shorter, is of the same width, and as in the 707, can easily be converted for the transport of either passengers or cargo, or both. The wings have very advanced high-lift devices for slow flying, triple-slotted trailing-edge flaps, and leading-edge flaps and slots but so cleverly retracted for high-speed flight as to be almost invisible.

PLATE 30. FLIGHT TODAY. THE BRITTEN-NORMAN ISLANDER

(By courtesy of Britten-Norman Ltd.)

This picture has purposely been placed between the F111 and the Boeing 727 on the one side, and the Jaguar and Concorde on the other. Why? Because among these sophisticated, complicated and fantastically expensive aircraft we have in the Islander a refreshing example of a return to the good old principles of aircraft design—simplicity and lightness, strength and accessibility, and, as aircraft go, even cheapness—a mere £20,000 or so as against hundreds of thousands or even a million or more; and a quite exceptional economy of operation. But the Islander, fixed undercarriage and all, is *not* an old-fashioned aeroplane, it is modern, it is an aircraft of today. It is a short-haul airliner with S.T.O.L. properties, a range of 700 miles with 10 people on board and a speed of about 160 m.p.h.; and with the versatility of a Boeing it can be changed in a matter of minutes from passengers to cargo, to ambulance, to executive comfort complete with cocktail cabinet. It is, too, an example of international co-operation in that the very British airframe is fitted with American equipment including engines, propellers, wheels and brakes, which have the great advantage of being interchangeable "off the shelf" all over the world.

PLATE 31. FLIGHT TODAY. THE ANGLO-FRENCH JAGUAR

(*By courtesy of the British Aircraft Corporation*)

Modern aircraft are so complicated, and so expensive, that not only are they beyond the means of individual firms such as existed before the war and are now amalgamated into groups or corporations, but they are beyond the financial resources even of nations. There has long been co-operation between Britain and the United States, though mostly in the form of one country building the aircraft and the other the engines or accessories; but recently a much bolder and more difficult attempt at collaboration has been made between Britain and France; and in spite of all the difficulties, political and technical, here in the Jaguar is practical evidence of a combined effort of BAC and Breguet Aviation. This is a military aircraft, but the example illustrated is a tandem two-seater advanced trainer which has the supersonic performance of its strike counterpart.

PLATE 32. FLIGHT TODAY. THE ANGLO-FRENCH CONCORDE

(By courtesy of the British Aircraft Corporation)

But here surely we have the most impressive example of international co-operation; the first supersonic airliner produced by the combined efforts of BAC and Sud-Aviation. There has been controversy about the need for supersonic transport at all, over the nuisance or damage to people and property from sonic booms, and, inevitably of course, over the enormous cost, but the fact remains that, at a time when Britain and France have found little on which to agree, their two aircraft industries have succeeded in this venture in spite of innumerable technical and other difficulties. Even now we do not know whether the Concorde will prove a commercial proposition, or whether it will come up to expectations in performance and other respects, but we do know that its mere production is an achievement of which Europe can be proud—and we wish it success. This photograph was taken on the 11th December, 1967, on the occasion of the roll-out of the first prototype, Concorde 001, at Toulouse.

Group 2: Plates 33–64

PLATE 33. LIGHTER THAN AIR. A CAPTIVE OR KITE BALLOON
(By courtesy of "Flight")

The kite balloon, as its name implies, has some of the properties of a kite and some of those of a balloon. Since the main object of flying is to get somewhere quickly, and since the lighter-than-air craft is most vulnerable, particularly in war, it is rather surprising to find that the kite balloon has proved more practically useful than the kite, the free balloon or the airship. For observation purposes, both in peace and war, and as a protective barrage for cities and shipping, the kite balloon was used in both World Wars—and still exists.

PLATE 34. LIGHTER THAN AIR. FREE BALLOONS
(*By courtesy of "Flight"*)

Free balloons, filled first with hot air then with hydrogen, enabled man to gain his first experiences in the air. It could hardly be called flying as we know it, but much was learnt about the atmosphere. Note the large size of the envelope required to lift even two or three men. There are still enthusiasts for ballooning—as a sport and as a vehicle for research. As recently as 1957 the balloon height record was raised to 101,516 ft, and in the winter of 1958–59 an attempt was made to cross the Atlantic in a balloon—all rather refreshing in an age of jets and space-ships.

PLATE 35. LIGHTER THAN AIR. AN AIRSHIP. THE GRAF ZEPPELIN

(*By courtesy of "Flight"*)

At one time, airships were thought by many people to be "natural" ships of the air. Why, then, has their history been so tragic? And why have they almost disappeared from the aeronautical scene? This picture gives the answer. Notice the bulky, vulnerable envelope, and compare it with the wings of any modern aircraft. But the airship is not quite dead yet, and new types are still being built in some countries.

PLATE 36. A FLOAT PLANE. THE FAIREY SWORDFISH

(*By courtesy of the Fairey Aviation Co. Ltd.*)

Although the float plane has, in the past, proved of great value and more than once has held the world's speed record, it, too, has now practically disappeared from the aeronautical scene. The reason for this may be summed up by saying that float planes are inferior to land planes in manoeuvrability and performance, and inferior to flying boats in their ability to withstand heavy seas.

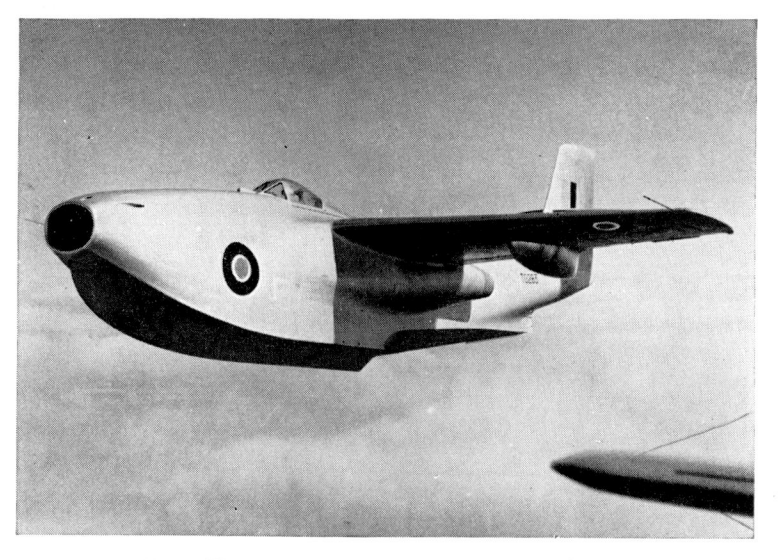

PLATE 37. A FLYING BOAT. THE SAUNDERS-ROE S.R.A.1

(*By courtesy of Saunders-Roe Ltd.*)

Although arguments for and against the airship were settled by its virtual disappearance, controversy continued to rage about the flying boat which at least survived to see the jet age. The whole question turns on whether, in view of the reliability and range of modern land planes, it is really necessary to be able to land on the sea.

PLATE 38. AN AUTOGIRO. THE CIERVA C.24

(*By courtesy of "Flight"*)

These strange looking aircraft, the autogiro and the helicopter, with their rotary wings, are in a class by themselves, and are now covered by the general term "rotorcraft". Although there were doubts about the future of these craft, the cynics were forced to admit that even the autogiro—which became a practical proposition before the helicopter—could do things which aeroplane designers had been trying in vain to do since the beginning of flying. Our chief interest in these two types is the difference between them. The autogiro has an engine and propeller which gives it forward motion while the wings are free to rotate.

PLATE 39. A HELICOPTER. THE WESTLAND W.G.13

(*By courtesy of Westland Helicopters Ltd.*)

In a helicopter, on the other hand, the wings are driven by the engine, and forward motion (or sideways or backwards) is achieved by tilting the axis of the rotating blades in the required direction. The majority of helicopters have been based on American design, but the Westland W.G.13 is all-British, and includes special features such as a simplified rotor head and gearbox, and plastic panels for parts of the structure—features aimed at increasing efficiency, reducing maintenance costs and increasing reliability. This anti-submarine version has all-weather capacity and a high-vertical-velocity non-rebound undercarriage enabling it to operate from small ships. The W.G.13 is powered by two turbine engines, and is essentially of utility type, suitable for the Services or as a civilian transport, and particularly for city centre to city centre operation.

Note the small propeller at the rear to prevent the aircraft rotating in the opposite direction to the main rotor, and to act as a rudder.

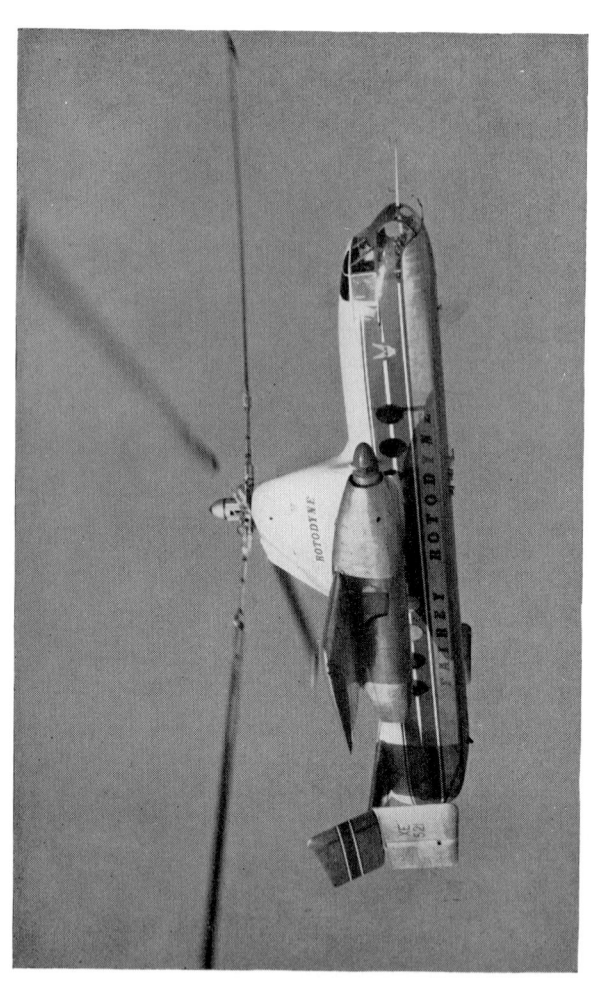

PLATE 40. HELICOPTER-CUM-AUTOGYRO. THE FAIREY ROTODYNE

(*By courtesy of the Fairey Aviation Co. Ltd.*)

And here is an even more interesting development, a helicopter and autogyro combined—and, indeed, a fixed-wing aircraft too. This was surely the first real V.T.O.L. airliner, but unfortunately it did not prove to be a commercial proposition. When flying as a helicopter—for take-off and landing—the four-blade rotor was powered by pressure jets at the tips (these can just be seen in the photograph). These jets were fed with compressed air from the two gas-turbine engines, which also provided the power, through the ordinary propellers, for cruising flight. The transition speed was from about 90 to 140 m.p.h.; below this speed most of the lift came from the power-driven rotor blades; above it the lift was shared between the fixed wings and the auto-rotating blades, the latter contributing some 55 per cent of the lift at high speeds.

PLATE 41. NEW TYPE ROTORCRAFT. THE WESTLAND WE-02
(*By courtesy of Westland Helicopters Ltd.*)

We have now seen three types of rotorcraft—the autogyro, the helicopter, and a combination of the two. They have achieved the much sought after vertical take-off and landing, but cannot rival the fixed-wing aircraft in speed or in lifting capacity for the power expended. But aircraft firms throughout the world are still experimenting—two or more contra-rotating rotors have been used (avoiding the need of the tail propeller). From the United States we hear of rigid rotors that do not have to twist and flap and attain much higher speeds in level flight, and of others that can even be folded out of the way in flight. But Westland have their own ideas of obtaining both V.T.O.L. and fixed-wing speed—instead of the more usual methods of having lifting jet engines, or vectoring the thrusts of jets downwards, or of tilting ordinary propellers or wings, they are experimenting with tilting helicopter blades of low disc loading, as seen in this model of the WE-02, which in full scale they expect to carry 80 passengers at speeds up to 380 m.p.h. over short ranges. To test the principle a smaller version, the WE-01, will be tried first.

PLATE 42. S.T.O.L., OLD STYLE. SCOTTISH AVIATION TWIN PIONEER

(*By courtesy of Scottish Aviation Ltd.*)

Although the autogyro and the helicopter, were generally accepted as one method of achieving slow or vertical take-offs and landings, they were not the only method, and there were many who believed that we should only achieve a speed range from zero up to really high speeds if, for the high-speed end, we used a more conventional fixed-wing type of aircraft. And so we went on experimenting with high-lift wings, and slots, and Fowler flaps and all that sort of thing—rather on the lines of the old Lysander (Plate 9). In this picture the Scottish Aviation Twin Pioneer, an outstanding S.T.O.L. type which first flew in 1955, is showing off its high-lift devices in a take-off from the apron at Munich Airport. But is this the answer? One cannot help but think back, or look back, to the Lysander—and wonder whether sufficient progress had been made in the intervening 20 years.

PLATE 43. V.T.O.L., OLD STYLE. THE SHORT S.C.1

(*By courtesy of Short Bros. and Harland Ltd.*)

Or is this the answer? Here is a Short research aircraft making its first free vertical take-off in 1958. In this we used what might be called the "flying bedstead" principle with four turbo-jet engines mounted vertically (with a limited degree of swivel) to give the thrust to lift the aircraft vertically—but in the attitude of normal horizontal flight. Yet it was not just a "flying bedstead," it was also a fixed-wing aircraft of conventional type, and moreover of conventional high-speed delta type, with yet another turbo-jet engine, of the same type and power as the others, mounted in the tail at 30° to the horizontal to give the forward speed. Its first conventional flight was made some 18 months before its first vertical flight, but the real problem with this arrangement is the transition from one kind of flight to another—in both directions of course. The ratio of one engine for forward flight to four engines for vertical flight, is a significant reminder of the principles of flight, lift/drag ratio, and so on. Although we have described it as "old style," the S.C.1 was still being used for experimental purposes in 1968.

PLATE 44. V./S.T.O.L., NEW STYLE. THE HARRIER
(*By courtesy of the Hawker Siddeley Group*)

But here surely we have signs of progress, for in the Hawker-Siddeley Harrier we have the nearest yet to an aircraft which can be used either for vertical or short take-off and landing, and also carry out at high speeds low-level reconnaissance and other missions from small dispersed areas, and so offer to an army immediate close support with bombs, rockets or cannons. Its remarkable take-off and landing properties, and equally impressive quick or slow changes from vertical to horizontal flight (Mach 1 has been exceeded), are obtained by "vectored thrust," i.e. by changing the direction of the thrust of some 20,000 lb of the Rolls Royce (Bristol) Pegasus engine. It can be refuelled in flight to increase the range of some 2,000 miles; and the rate of climb to 10,000 ft is quite exceptional. As it is essentially a military type, a more detailed description of its remarkable capability and performance cannot be given for reasons of security. A two-seater training version is being produced.

PLATE 45. V.T.O. THE SATURN V MOON ROCKET

(*By courtesy of the Boeing Company*)

Perhaps the time has come when the sub-title of this book, "How and Why an Aeroplane Flies," should be changed, because by no stretch of imagination (and still less by dictionary definition) could this be called an aeroplane, and the same applies to other illustrations in this second group of photographs—yet surely they come under the general heading of "flight." In 1903 it was all we could do to make an engine with sufficient power to propel an aeroplane (remembering that this requires much less thrust than the weight); then it took another 30 or 40 years to produce an engine to give enough thrust to lift a helicopter vertically; but here in the Saturn V, the full 3-stage rocket 364 ft tall, weighs more than 6 million pounds, and is capable of carrying 45 tons of "payload" to the moon. The booster with five engines, gives a lift-off of $7\frac{1}{2}$ million pounds, the second stage provides 1 million pounds of thrust, and the third a mere 230,000 pounds for orbital operations. This is vertical take-off indeed.

PLATE 46. CONTRASTS. A SAILPLANE. THE SLINGSBY SKYLARK 4

(*By courtesy of Slingsby Sailplanes Ltd.*)

What a contrast to the moon rocket of the last plate or even the more modest Bristol rocket on the opposite page! But this is flight under ideal conditions; no brute force from an engine, no thrust except the force of gravity, no vertical lift except from atmospheric currents and the skill of the pilot—but clean design, high aspect ratio, everything in fact to give maximum lift with minimum drag, minimum weight, and wonderful, peaceful, silent flight—surely flight without formulae put into practice. But of course there are limitations when it is a question of getting from A to B, still more to the moon! So efficient is the design that it is difficult to stop flying, and spoilers, or air brakes, are needed to spoil the lift/drag ratio and so steepen the gliding angle and take off surplus speed before landing.

PLATE 47. CONTRASTS. A ROCKET. THE BLACK KNIGHT
(By courtesy of Westland Aircraft Ltd.)

Black Knight is a research rocket designed to investigate the problems of re-entry into the earth's atmosphere. It is a co-operative effort, conceived at Farnborough, constructed by Westland (Saunders-Roe), and powered by Bristol Siddeley. It is fired from the Woomera range in Australia, which has the great advantage that re-entry parts can be recovered on the ground and easily reached from the range. To ensure that the test vehicle comes down within reasonable distance, a second stage is released as soon as the first is burnt out (at about 350,000 ft), and is made to spin and descend vertically until before real re-entry the second-stage rocket is fired to give suitable re-entry velocity at the most interesting stage between 200,000 and 100,000 ft. The project and the data obtained are now shared between Britain, Australia and the United States. A satellite launch vehicle, Black Arrow, has now been designed based on the experience with Black Knight.

PLATE 48. CONTRASTS. A HOVERCRAFT. THE B.H.C. SRN.6

(*By courtesy of Westland Aircraft Ltd.*)

The hovercraft, too, has no claim to be an aeroplane, and perhaps even less than the rocket to be capable of "flight." But it is designed and built on aircraft principles, usually driven by aircraft engines and propellers, and constructed by aircraft firms. As to whether it flies, it does at least travel through the air, without contact with land or water, resting on a cushion of air created by power-driven fans instead of on wings as in an aeroplane. It has great versatility, can travel at speeds up to 60 knots over shallow waters, mud, marsh, sand, rivers and rapids, and ice and snow as in this illustration of trials in Sweden. It can be used as a passenger ferry, for crash rescue purposes, as a mobile medical centre, even as a form of ship-based transport. It can cope with waves up to 4 or 5 ft, but at reduced speed. A large version, the 165 ton SRN.4, can carry some 500 passengers, or cars or other freight, at more than 70 knots over opensea ferry routes such as the English Channel. As a British invention (at any rate in a practical form), it is unusual in that it has been first built and developed in Great Britain.

PLATE 49. CONTRASTS. A HYDROFOIL CRAFT. GIBBS & COX SEA LEGS

(By courtesy of Gibbs and Cox Inc., New York.)

Still less an aeroplane? Still less capable of "flight"? Perhaps, but as explained on page 31 the hydrofoil craft comes very near to the definition of an aeroplane on page 2 if one merely substitutes "driven through the water" for "driven through the air." Moreover, while it is true that some types of hydrofoil craft travel as near the surface of the water as do hovercraft, the "high-flyers" fly well clear of the water (except of course the submerged foils), and can claim a smoother passage through, rather than over, higher waves than hovercraft of comparable size. The essential difference, and disadvantage, of the hydrofoil craft is that it is not amphibious; it can only travel over water. Another difference is that, although there have been a few dedicated British inventors, they have received little or no encouragement compared with nearly every other maritime nation in the world. There has been a recent revival of interest in hydrofoils for use on seaplanes and flying boats, an idea that goes back to the first war.

PLATE 50. TRAINER, OLD STYLE. THE AVRO 504K
(By courtesy of "Flight")

The most famous training aeroplane of the early days and perhaps of all time. Conventional in those days except for its undercarriage (a combination of wheels and skid) and the absence of a fin. Its excellent training qualities were due to a kind of neutral stability; you had to fly it, and yet it was easy to fly.

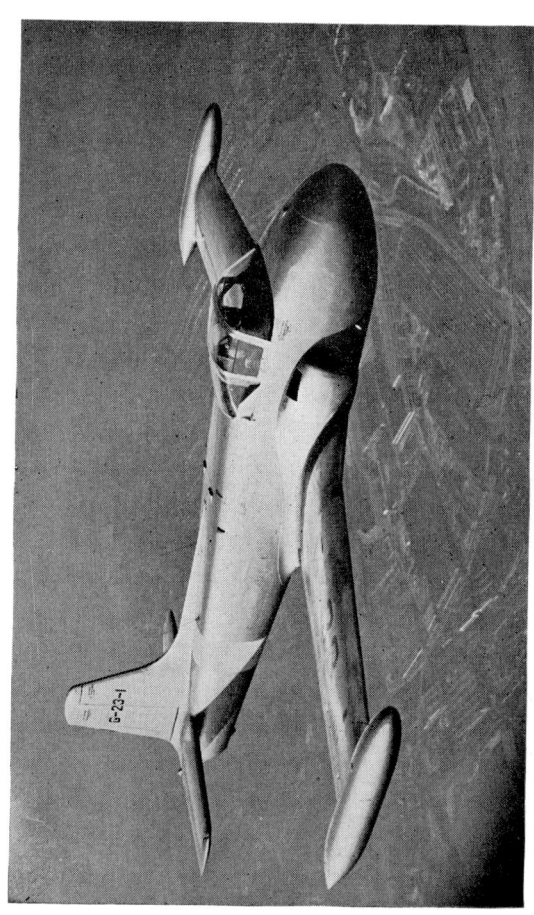

PLATE 51. TRAINER, NEW STYLE. THE HUNTING JET PROVOST

(*By courtesy of Hunting Aircraft Ltd.*)

This was the world's first military jet training aircraft in which pupil pilots learnt to fly from the beginning of their flying career—it is also interesting because its qualities as a trainer were first tried out in a piston-engine form. It has side-by-side seats and full dual controls and, in its jet form, is powered by an Armstrong Siddeley "Viper" turbo-jet engine.

PLATE 52. PUSHER, OLD STYLE. THE MAURICE FARMAN SHORTHORN

(By courtesy of "Flight")

This picture does indeed tell the story of the early days of flying. Notice the engine with its large-diameter, two-bladed, pusher propeller just behind the pilot's back. In order to give space for the propeller it was necessary to support the tail on booms instead of having a fuselage. There was no dual control as we know it, and the instructor sat close behind his pupil to be able to control his arm and leg movements, and shout his instructions, which could often be heard from the ground. The author had his first lessons in the Longhorn, which had a front elevator and even more struts and wires.

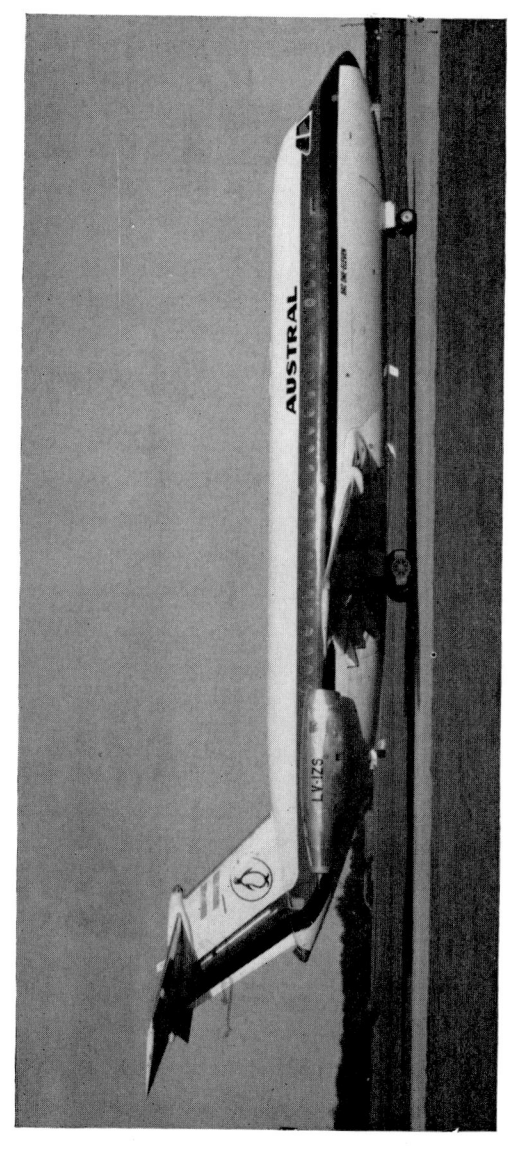

PLATE 53. PUSHER, NEW STYLE. THE BAC ONE-ELEVEN

(*By courtesy of the British Aircraft Corporation*)

This excellent aircraft has already been described under Plate 27. This particular model, which has been specially adapted for service in South America, carries 74 passengers at cruising speeds up to 550 m.p.h. The picture shows clearly one of the two rear-mounted Rolls Royce Spey turbo-fan engines (the pusher arrangement of today).

PLATE 54. PARASITE DRAG. THE VICKERS VIRGINIA
(*By courtesy of "Flight"*)

These two pictures show a contrast both in time and knowledge and are certainly not a reflection on the designers of the Vickers Virginia, a type of aircraft that had long and honourable service to its credit, and which was a descendant of the Vimy, the first aeroplane to make a non-stop flight across the Atlantic—in 1919. This is what people at that time expected a large aeroplane to look like.

PLATE 55. STREAMLINING. THE ENGLISH ELECTRIC CANBERRA
(By courtesy of the British Aircraft Corporation)

The Canberra is hardly a new type of aircraft—it has been flying for nearly 20 years—but as a British long-distance strike/reconnaissance aircraft it is still going strong, and since the cancellation first of the TSR2 (Plate 62), then the Anglo-French swing-wing project, then the F111K, each of which should have replaced it for the Royal Air Force, it is likely to go on for some time yet. But it has been chosen for this story in pictures as a fitting contrast to the Virginia, in that whereas the Vimy made the first direct flight across the Atlantic (in 16½ hours), this particular Canberra as long ago as 1952 crossed the Atlantic both ways during daylight in one day—for the first time by any aircraft—and then held three records, the easterly, westerly and two-way trips—the easterly in 3hr 25min at an average speed of 605 m.p.h. Its clean lines are clear enough, but its performance is an even better proof of its efficiency.

PLATE 56. THRUST FOR FLIGHT. PROPELLER AND ROTARY ENGINE

(By courtesy of the Bristol Aeroplane Co. Ltd.)

Before its time? The Bristol Scout, only military machine in production at the outbreak of the First World War in 1914, and the forerunner of the other war-time scouts and fighter aeroplanes. Very typical of the wire-braced biplane construction of conventional type at that time. Fitted with an 80 hp Gnome rotary engine and fixed-pitch propeller.

The petrol-air mixture was fed into the crankcase and sucked into the cylinders through automatic inlet valves, but the strangest feature of the engine was that there was no proper throttle control, and in order to come in to land, one had to switch the engine on and off by a "blip" switch on the control column.

PLATE 57. THRUST FOR FLIGHT. TURBO-PROPS

(*By courtesy of Vickers-Armstrongs (Aircraft) Ltd.*)

As has been explained in the text, there are three possible methods of propulsion for modern aircraft—propellers, jets and rockets. Generally speaking propeller propulsion is most suitable, and economical, for subsonic speeds; jet propulsion for high subsonic, transonic and moderate supersonic speeds; rocket propulsion for even higher speeds—but rocket propulsion entails carrying so much fuel, and is so extravagant in fuel, that for manned aircraft it is used mainly to act as a boost to other methods. This picture shows typical modern propellers—as fitted to a Vickers Viscount. But the engines as is the usual modern practice for high-powered aircraft, are internal combustion turbines very similar to those used for jet propulsion; and, in fact, there is an element of jet propulsion from the turbine exhaust or jet pipe. So this is called propulsion by turbo-props as distinct from turbo-jets. The engines in the illustration are Rolls Royce Darts, three-stage turbines, each giving at take-off some 1,800 hp in addition to a thrust from the jet of some 500 lb. Modern propellers may not look very different from old types, but this description (of the 16 ft diameter propellers on a Britannia) gives some idea of the refinements which have been introduced—"constant-speeding, feathering, reversing, hydraulic pitch locking and automatic r.p.m. synchronization."

PLATE 58. THRUST FOR FLIGHT. TURBO-JETS AND ROCKET BOOSTERS

(*By courtesy of the Boeing Airplane Company and "The Aeroplane"*)

Here is a striking example of jet and rocket propulsion combined—the Boeing "Stratojet" using some of its eighteen rockets as boosters to give extra power for take-off. It should be noted that the "power" of a turbo-jet or rocket engine is not given in horse-power, because horse-power involves the moving of a certain force through a certain distance in a certain time; this can be applied to an engine turning a propeller even when the aircraft is not moving forward, but whatever the thrust given by a pure jet or rocket the horse-power would be nil if there were no forward movement. So the effectiveness of a jet or rocket engine is judged by its static thrust which may be anything up to about 20,000 lb for a turbo-jet, and up to about 10,000 lb for a rocket engine.

PLATE 59. THRUST FOR FLIGHT. ONE MAN-POWER

(By courtesy of the Man-Powered Aircraft Group at the University of Southampton)

We can send rockets to the moon and Venus and outer space; we can fly at over 4,000 miles an hour; but one thing we still cannot do is to fly under our own power like even the most modest of the birds of the air. But here at least is a gallant attempt, successful one might almost say except that it did not come up to the simple requirements to win the various prizes that have been offered for the first real flight made by man-power alone. We live in a strange world in that such a feat seems to require more brains, more skill in design, more research and development of materials, and certainly more physical strength and determination than such a colossal enterprise as the Concorde—but it is certainly cheaper!

PLATE 60. PUTTING THE BRAKES ON. THE ENGLISH ELECTRIC LIGHTNING

(By courtesy of the English Electric Co. Ltd.)

Everyone who drives on the roads knows—or should know—that it takes a great deal more than twice the distance to pull up from 60 m.p.h. than from 30 m.p.h. But what about putting the brakes on in a vehicle that is capable of 1,200 m.p.h.? Well, in an aircraft it is done in two stages—first in the air, then on the ground. Here we see the first stage as applied to the Lightning. Flaps are fully down, increasing the drag and so lowering the speed, but at the same time increasing the lift and so making it safe to fly more slowly. The air brakes are fully on (the starboard brake, sticking out against the airflow, can be seen just in front of the tail fin). Then, not to be despised as drag increasers, are the wheels and the undercarriage legs, now lowered for landing. These devices between them, together with the increase of angle of attack at the will of the pilot, can reduce the speed of this supersonic aircraft to a mere 100 m.p.h. or so, and thus give the remarkable speed range of more than 10 to 1. But, once on the ground, even 100 m.p.h. is a formidable speed, and, in addition to powerful wheel brakes, a parachute is trailed behind; and on some aircraft, reversible propellers or reversible jets are used.

PLATE 61. PUTTING THE BRAKES ON. THE HANDLEY PAGE VICTOR
(By courtesy of "Flight")

This view of the under surfaces of a Handley Page Victor reveals the general tendency in the design of aircraft for transonic speeds. A special feature of the Victor is the crescent-shaped leading edge of the wing, which is similar to the Valiant and was referred to under Plate 20. But the most interesting feature of this photograph of the Victor is that it is about to land and is displaying all its braking and slow-speed devices rather like the Lightning in Plate 60. Notice the air brakes at the rear end of the fuselage, the lowered leading edge flaps increasing the camber, the Fowler flaps at the trailing edge increasing both camber and area, and the lowered wheels and undercarriage legs. But effective as these devices are for their purpose, they cannot altogether disguise the clean lines of the Victor.

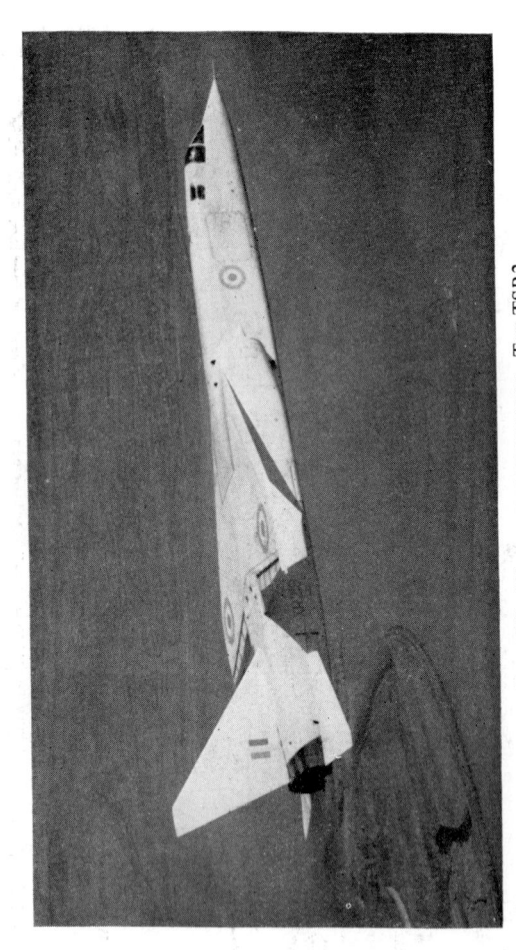

PLATE 62. WHAT MIGHT HAVE BEEN. THE TSR2

(*By courtesy of the British Aircraft Corporation*)

It will never be known what the British aircraft industry might have achieved in the 1970s if they had been given the resources of the corresponding industries in the United States and in the Soviet Union. Among other things they would have produced the TSR2 not just a new aeroplane, but a complete weapon system (the initials TSR standing for "tactical strike and reconnaissance") capable of penetrating 1,000 miles into enemy territory in all weathers by day or night. Space does not permit even a mention of all the advanced features of this aircraft; the thin delta wing of special shape little affected by gusts—even at Mach 2; large blown flaps giving a tendency towards S.T.O.L. performance; the anhedral at the tips to avoid interference of vortices with the tail; no ailerons, no fences, sawcuts or vortex generators; an all-moving slab fin and a rolling tail-plane; two turbo-jets each of some 33,000 lb thrust; and much else too. No wonder that in 1963 the technical editor of *Flight* wrote: "This new all-British aircraft is one item of equipment we cannot afford not to have." But it was not to be—we could not afford to have it—we ordered the American F111 instead—and then we could not afford to have that either.

PLATE 63. ANOTHER "MIGHT HAVE BEEN"? THE BOEING S.S.T.

(*By courtesy of the Boeing Company*)

Whereas the prototype of the Anglo-French S.S.T. Concorde is actually in existence, the Americans are in the unusual position of being behind, partly because they held a competition between different firms—and this took time—and partly because they are aiming at a higher Mach number, in fact a cruising speed of at least 1,800 m.p.h. The Boeing design, which at first was the one chosen, employed the swing-wing principle, and this multiple-exposure photograph of a model shows the low, medium and high-speed positions of the wings. This aircraft was designed to carry more than 200 passengers over intercontinental distances at supersonic speeds—perhaps it is a symbol of the future? Since this picture was taken the design was modified by increasing the length by 12 ft, and by fitting a small canard surface (a "tail" at the front) to improve longitudinal control. But, after all this, the swing wing principle was abandoned, so the Americans were even further behind, in fact back to square one.

PLATE 64. FIRST FLIGHT OF CONCORDE 002, 9TH APRIL, 1969

(*By courtesy of the British Aircraft Corporation*)

It is perhaps fitting to end the second group, as the first, of this story in pictures with a photograph of the Concorde—but this time in actual flight.

The French version, Concorde 001, first flew on 2nd March, 1969, at Toulouse in France, and this was soon followed by the British version, 002, which flew from Filton, Bristol, and landed at R.A.F. Fairford after a perfect flight. In both instances the first flights were made without retracting the undercarriage and with the nose in the drooped position (as in the illustration) to improve the pilot's visibility for take-off and landing—in later flights the undercarriage was retracted and the nose raised, thus giving a completely different impression of the clean lines of this remarkable aircraft.

Whatever the controversy over this project, it is clearly a landmark, both in technical achievement in the design of aircraft and in technological co-operation between nations—the beginning of a new chapter in the history of civil aviation.